T0003539

Springer Undergraduate Texts in Mathematics and Technology

Springer Undergraduate Texts in Mathematics and Technology (SUMAT) publishes textbooks aimed primarily at the undergraduate. Each text is designed principally for students who are considering careers either in the mathematical sciences or in technology-based areas such as engineering, finance, information technology and computer science, bioscience and medicine, optimization or industry. Texts aim to be accessible introductions to a wide range of core mathematical disciplines and their practical, real-world applications; and are fashioned both for course use and for independent study.

More information about this series at http://www.springer.com/series/7438

Giray Ökten

Probability and Simulation

 Springer

Giray Ökten
Department of Mathematics
Florida State University
Tallahassee, FL, USA

ISSN 1867-5506   ISSN 1867-5514   (electronic)
Springer Undergraduate Texts in Mathematics and Technology
ISBN 978-3-030-56069-0   ISBN 978-3-030-56070-6   (eBook)
https://doi.org/10.1007/978-3-030-56070-6

Mathematics Subject Classification: 97K50, 60-01

This Springer imprint is published by the registered company Springer Nature Switzerland AG
The registered company address is: Gewerbestrasse 11, 6330 Cham, Switzerland

To my mother Fikret, and my daughter Arya

Preface

The content of this book is mostly based on an upper-level undergraduate math class on stochastic models and computing that I taught at Florida State University. The style of delivery took shape when I started teaching the content for the Young Scholars Program (YSP) at Florida State University. YSP is a residential summer program for Florida high school students interested in STEM, and most of the students are rising seniors. All students have studied calculus, and some probability, before attending this program. The YSP classes were three times a week, for six weeks, and each class meeting lasted for two hours. Students were very motivated and had strong backgrounds, but this was probably their first class in college. These factors created a unique set of challenges and opportunities for me as an educator. I decided to develop an inquiry-based learning course based on the material I had previously taught to upper-level math majors.

Each chapter of the book introduces a topic that can be covered in 50–60 minutes, followed by a significant project that teams of students can work on for about 40 minutes. The role of the instructor in the second part of the class is that of a facilitator. The challenge for me was to find projects that would excite students, and to cover the necessary content in the first half of the lecture. Some projects students did not like were removed, and some others modified based on the feedback I got in the classroom.

At the core of the content lie the traditional topics: probability axioms leading to Bayes' theorem, discrete and continuous random variables, Markov chains, and Brownian motion. The layer surrounding the core is the applications, and the book features topics that are not commonly treated in standard undergraduate textbooks, such as randomized algorithms, randomized surveys, Benford's law, and Monte Carlo methods. Some of these applications are covered in the lectures, some in the projects.

The pedagogical approach used in the book has three main elements:

- Games: I use this word in a general sense. In some projects, students flip coins, play dice games, shoot a Nerf gun, or act out the Monty Hall problem. In other projects, they split into groups and design a randomized survey with a sensitive question and interview each other, look up some data on the Internet and try to guess the distribution of its digits, or try to generate random numbers collectively. Any social activity that changes the traditional style of lecturing I call a game. Some of the games I borrowed from the literature help students with the solution to the problem in hand, such as the coupon collector's problem (Project 6).

- Data: Most projects involve data. In some cases, data is generated by students by flipping coins or rolling dice. In some cases it is downloaded and analyzed. The use of data serves many purposes. Sometimes, we use it to get an insight into a problem, sometimes to understand how it can be modeled by a random variable.

- Simulation: I used simulation as a tool to understand probability concepts and as a computational tool. Simulation and random numbers are used to estimate probabilities and expectations. Histograms of random numbers and data are used to motivate the concept of a probability density function and the meaning of a good probability model. After several examples of using qualitative statements to suggest data follows a specific distribution, a formal treatment is discussed via the χ^2-test in Chapter 3.

The book uses the Julia programming language (version 1.1.0). The Julia software is free on a license from MIT. The simplicity of Julia allows incorporating the computer code within the main narrative, without a significant distraction to the main flow of the narrative. A tutorial on Julia can be found in Chapter 1 of *First Semester in Numerical Analysis with Julia*, an open-access textbook I wrote on numerical analysis, published by Florida State University Libraries.[1] For readers who might prefer the Python programming language, a companion book *Probability and Simulation: A Python Companion* cowritten with Yaning Liu is available on the Springer web page for the book. The companion book has the Python translations of the Julia codes presented here.

One of my goals in writing this book and teaching the class was to utilize a broad range of activities: playing games to gain insight, running simulations to strengthen intuition, and proving theorems. Part of the beauty of probability is how it is possible to implement these different strategies.

I thank my colleague Brian Ewald who has co-taught the YSP class with me and suggested numerous clarifications and corrections. I thank Yoshi Fujiwara who shared the data used in Example 3.3. I thank Arya Ökten who drew the cartoons. Finally I thank the YSP students whose genuine enthusiasm provided me with the motivation to complete this book.

Tallahassee, FL, USA   Giray Ökten
June 2020

---

[1]https://doi.org/10.33009/jul.

Contents

Chapter 1
Probability

If shapes such as lines and triangles are emblematic of geometry, and symbols such as x, y are emblematic of elementary algebra, then the fair die or coin is the symbol of probability. When we flip an unbiased coin, we know the outcome will be either heads or tails, with equal likelihood. The concept of a **random experiment** is an abstraction of this phenomenon: it is an experiment which has several possible outcomes, and we cannot tell what the outcome will be a priori.

1.1 Axioms of probability

The precise definition of **probability** and **probability space** is given next. There are three important concepts introduced in the definition, and to help understand these concepts, we will discuss an example within the definition.

Definition 1.1 (Probability space)   A probability space consists of

- A sample space Ω, which is a collection of all possible outcomes of a random experiment;

  - *If the random experiment is rolling a fair die, then the sample space would be $\Omega = \{1, 2, 3, 4, 5, 6\}$.*

- A family of sets $\mathcal{F}$ representing the allowable events (each set in $\mathcal{F}$ is a subset of Ω);

  - *For the fair die example, any subset of $\Omega = \{1, 2, 3, 4, 5, 6\}$ is an allowable event; these are events for which we can assign a probability.*

- A probability function $P : \mathcal{F} \to \mathbb{R}$ that satisfies the following conditions:

  1. for any event E, $0 \le P(E) \le 1$
  2. $P(\Omega) = 1$

© The Editor(s) (if applicable) and The Author(s), under exclusive
license to Springer Nature Switzerland AG 2020
G. Ökten, *Probability and Simulation*, Springer Undergraduate Texts
in Mathematics and Technology, https://doi.org/10.1007/978-3-030-56070-6_1

3. if $E_1, E_2, \ldots$, is any finite or countably infinite[1] sequence of pair-wise mutually disjoint events (which means, any two distinct events E_i, E_j from the
sequence has empty intersection) then

$$P(\bigcup_{i \geq 1} E_i) = \sum_{i \geq 1} P(E_i).$$

We say $P(E)$ is the probability of the event E.

– *For the fair die example, a probability function can be defined as follows: if
E is an event, then set*

$$P(E) = \frac{number\ of\ outcomes\ in\ E}{6}.$$

For instance, if E is the event that the outcome is even, then $E = \{2, 4, 6\}$
and $P(E) = 3/6$. *It is not too difficult to show that this function satisfies the
conditions listed in the definition.*

Definition 1.2 (Discrete and continuous spaces) We say the probability space is
discrete if Ω is finite or countable[2] and $\mathcal{F}$ is all subsets of Ω. The probability space
is continuous if Ω is uncountable—how $\mathcal{F}$ is defined varies.

Note 1.1 The probability function satisfies the following identities.

• $P(E \cup F) = P(E) + P(F) - P(E \cap F)$
• $P(\cup_{i \geq 1} E_i) \leq \sum_{i \geq 1} P(E_i)$
• if $E \subseteq F$, then $P(E) \leq P(F)$
• $P(E^c) = 1 - P(E)$ where E^c is the set complement of E

Remark 1.1 Sometimes we use a comma for the intersection symbol and write
$P(E, F)$ for $P(E \cap F)$.

---

[1] A set is countably infinite if we can *count* its elements, which means labeling its elements using the
natural numbers as 1, 2, ... Sets whose elements cannot be labeled as such are called *uncountable*
sets. An example of an uncountable set is the set of real numbers $\mathbb{R}$.

[2] A set is countable if it is finite or countably infinite.

1.2 Verifying polynomial identities

Suppose we want to check whether the below equality is correct:

$$(x + 1)(x + 2)(x + 3)(x + 4) = x^4 + 10x^3 + 35x^2 + 50x + 26.$$

We can follow two approaches:

1. Multiply out the left-hand side, collect like terms, and check if it is equal to the right-hand side;
2. Substitute $x = 0$ in both sides of the equation to get $24 = 26$, and conclude the equality is wrong.

In general, suppose we want to verify the identity

$$F(x) \equiv G(x) \tag{1.1}$$

where the polynomials F, G are written in different forms:

$$F(x) = \prod_{i=1}^{d} (x - a_i) \text{ and } G(x) = \sum_{i=0}^{d} c_i x^i.$$

The straightforward approach is to multiply out the terms in $F(x)$, collect the like terms, and verify the identity. The cost of this operation is $O(d^2)$ multiplications of coefficients: this means the number of multiplications grows like Cd^2 where C is some positive constant. Next we will discuss another approach: an approach which is computationally cheaper, but does not always give the right answer!

1.2.1 Randomized algorithm

Follow the steps below to test the polynomial identity Eq. (1.1):

- Pick an integer r, uniformly at random from $\{1, 2, \ldots, 100d\}$ where d is the degree of the polynomial.
- Compute $F(r), G(r)$.
- If $F(r) \neq G(r)$, declare the two polynomials are not equivalent.
  If $F(r) = G(r)$, declare the two polynomials are equivalent.

We call this a randomized algorithm since it uses random numbers in its execution. Let's see why the algorithm does not always give the correct answer. If the algorithm declares the polynomials are not equivalent, it is always correct. However, when the algorithm declares the two polynomials are equivalent, there is a chance it is giving the wrong answer. We want to find this probability.

Precisely, we want to find the probability that for some integer r, $F(r) = G(r)$, but $F(x) \not\equiv G(x)$. With this assumption, observe that $F(x) - G(x)$ is a nonzero polynomial of degree (at most) d, and $F(r) - G(r) = 0$. In other words, r is a root of the polynomial $F(x) - G(x)$.

Then the question becomes, what is the probability that a randomly picked integer r from $\{1, 2, \ldots, 100d\}$ happens to be a root of $F(x) - G(x)$? The polynomial $F(x) - G(x)$ has at most d distinct roots. How many of these roots will lie in $\{1, 2, \ldots, 100d\}$? At most d. Then the probability in question is at most $\frac{d}{100d} = \frac{1}{100}$.

What about the cost of the randomized algorithm? Notice that evaluating $F(r)$ (or $G(r)$) requires d multiplications. Therefore the cost is $O(d)$ multiplications.

Going back to our randomized algorithm, let's define E as the event that the algorithm fails to give the correct answer. We showed that

$$P(E) \leq \frac{1}{100}.$$

Another way to state this result is that the algorithm gives the *correct answer* with a chance of at least 99%.

Can we improve this probability? There are two ways we can go about it.

1. Increase the sample space: if we pick r at random from $\{1, 2, \ldots, 1000d\}$, then the probability of failure will be at most $\frac{1}{1000}$.
2. Run the algorithm multiple times:

   - Pick numbers $r_1, r_2, \ldots, r_k$ at random, and independently, from the set $\{1, 2, \ldots, 100d\}$
   - Compute $F(r_1), G(r_1); F(r_2), G(r_2); \ldots ; F(r_k), G(r_k)$
   - – If $F(r_i) \neq G(r_i)$, for some $i = 1, \ldots, k$, declare the two polynomials are not equivalent.

      – If $F(r_1) = G(r_1), F(r_2) = G(r_2), \ldots, F(r_k) = G(r_k)$, declare the two polynomials are equivalent.

   In this new setup, the probability that the algorithm gives a wrong answer is the probability that $F(r_1) = G(r_1), F(r_2) = G(r_2), \ldots, F(r_k) = G(r_k)$ but $F(x) \not\equiv G(x)$.

   Let E_i be the event that r_i is a root of $F(x) - G(x)$, in other words, the event that corresponds to $F(r_i) = G(r_i)$. Then, as before,

$$P(E_i) \leq \frac{d}{100d} = \frac{1}{100}.$$

However, what we really want is the following probability:

$$P\left(F(r_1) = G(r_1), F(r_2) = G(r_2), \ldots, F(r_k) = G(r_k)\right) = P(E_1, E_2, \ldots, E_k).$$

We need the following definition to compute this probability.

Definition 1.3 (Independent events) Two events E and F are independent if

$$P(E \cap F) = P(E)P(F).$$

In general, events $E_1, E_2, \ldots, E_k$ are mutually independent if

$$P(\cap_{i \in I} E_i) = \prod_{i \in I} P(E_i)$$

for any subset $I \subseteq \{1, 2, \ldots, k\}$.

In our example, the sets E_i are independent, since r_i are independently generated random numbers. Then

$$P(E_1, E_2, \ldots, E_k) = \prod_{i=1}^{k} P(E_i) \leq \left(\frac{1}{100} \right)^k.$$

Therefore, when we run the algorithm k times using independent random numbers, the probability that the algorithm fails to give the correct answer is less than $(0.01)^k$.

1.3 Project 1: Verifying identities using Julia

1. Write a Julia function [3] called **verifyid** such that

   - The function takes three inputs: the polynomials F, G, and the degree of F (and G), call it d.
   - The output is one of the statements: "the polynomials are equivalent" or "the polynomials are not equivalent".

   Here are more details and hints for the code:

   - Initialize a Boolean variable **flag** as flag = true, and initialize $n = 1$.
   - Write a **while** statement that does the following while flag = true and $n \leq 10$:
     - generate a random integer r between 1 and $100d$; this can be done by $r = rand(1 : 100 * d)$
     - set **flag** to the truth value of $F(r) == G(r)$
     - increment $n = n + 1$
   - After the **while** statement is executed, print "the polynomials are equivalent" if flag = true. Otherwise print "the polynomials are not equivalent".

2. What is the probability that the function **verifyid** returns "the polynomials are equivalent" when in fact they are not equivalent?

3. Take turns to test your team member's **verifyid** code as follows:

   a. Make up a product, such as $F(x) = (x - 5)(x - 10)(x + 3)(x - 2)(x + 25)$.
   b. Type this in WolframAlpha to find an equivalent expression in standard form. For the above, it is $G(x) = x^5 + 11x^4 - 321x^3 + 865x^2 + 3200x - 7500$.
   c. You have two options: either report F and G as they are to your teammate, or make a little change in G and report it.
   d. Your teammate will check if F and G are equivalent or not.

---

[3] See Chapter 1 of Ökten[19] for a tutorial on Julia.

1.4 Verifying matrix multiplication: Freivalds' algorithm

The problem we want to solve is to verify the equality $AB = C$ where A, B, C are square matrices of size n. The obvious approach to verify $AB = C$ is to simply multiply the matrices A and B and then check if the result matches C. This approach is sound, but it can be costly if the matrices are very large. The cost of matrix multiplication is $O(n^3)$. (There are sophisticated algorithms that bring the cost down to $O(n^{2.37})$.)

Here we discuss a randomized algorithm, Freivalds' algorithm [8], to answer this question. These are the steps of the algorithm:

- Generate a uniform random column vector $r = [r_1, r_2, \ldots, r_n]^T \in \{0, 1\}^n$ by independently generating n uniform random numbers $r_i, i = 1, 2, \ldots, n$ from $\{0, 1\}$.
- Compute $A(Br)$ and Cr.
- If $A(Br) \neq Cr$, declare $AB \neq C$. If $A(Br) = Cr$, declare $AB = C$.

Clearly, if the algorithm says $AB \neq C$, it is always correct. However, if the algorithm says $AB = C$, then there is a possibility of a wrong conclusion: this happens when we were unlucky to find a vector r such that $A(Br) = Cr$, even though $AB \neq C$. (An example of such an unlucky choice would be $r = [0, \ldots, 0]^T$.)

We want to know the probability of this algorithm giving an incorrect answer. However, before we do that, let's ask whether this algorithm will be any faster than simply multiplying the matrices. Observe that multiplying a matrix by a vector takes $O(n^2)$ multiplications and Freivalds' algorithm has two such multiplications. This will be much smaller than $O(n^3)$ as n gets larger.

Theorem 1.1 *If $AB \neq C$ and if $r = [r_1, r_2, \ldots, r_n]^T$ is chosen uniformly at random from $\{0, 1\}^n$, then*

$$P(ABr = Cr) \leq \frac{1}{2}.$$

Proof Let $D = AB - C$. The hypothesis is D is a nonzero matrix. We want to find the probability that $Dr = 0$. Let's denote the rows of D as $d_1, d_2, \ldots, d_n$. We have

$$[D] \begin{bmatrix} r_1 \\ r_2 \\ \vdots \\ r_n \end{bmatrix} = \begin{bmatrix} d_1 \cdot r \\ d_2 \cdot r \\ \vdots \\ d_n \cdot r \end{bmatrix} = \begin{bmatrix} 0 \\ 0 \\ \vdots \\ 0 \end{bmatrix},$$

where the notation $d_i \cdot r$ means the dot product of the vectors d_i and r. Since D is a nonzero matrix, one of its row vectors should be a nonzero vector. Without loss of any generality, assume $d_1 \neq 0$ (here 0 means the zero vector whose components are all equal to 0). From the above equation, we must have

$$d_1 \cdot r = \sum_{j=1}^{n} d_{1j} r_j = 0 \qquad (1.2)$$

(as well as the other products) where $d_{11}, d_{12}, \ldots, d_{1n}$ are the components of the row vector d_1. Since $d_1 \neq 0$, one of the d_{1j} must be nonzero. Without loss of any generality, let's assume d_{11} is not zero. Then we can solve Eq. (1.2) for r_1 to get

$$r_1 = -\frac{\sum_{j=2}^{n} d_{1j} r_j}{d_{11}}. \qquad (1.3)$$

Here is the crucial idea of the proof: imagine generating the components of the random vector r in the following order: $r_n, r_{n-1}, \ldots, r_2, r_1$. Once r_2 is generated, the right-hand side of Eq. (1.3) has a set value: if this value is 0 or 1, then r_1, which is the random number (0 or 1) generated next, will be equal to the value with probability 1/2. If the value is not 0 or 1, then Eq. (1.3) will not hold, and thus $d_1 \cdot r \neq 0$. Therefore the probability that Eq. (1.3) holds is at most 1/2, or equivalently, $P(d_1 \cdot r = 0) \leq 1/2$. Finally, note that

$$P(Dr = 0) = P(d_1 \cdot r = 0, d_2 \cdot r = 0, \ldots, d_n \cdot r = 0) \leq P(d_1 \cdot r = 0) \leq 1/2.$$

Example 1.1 Let $A = \begin{bmatrix} 2 & 3 \\ 1 & 4 \end{bmatrix}$ and $B = \begin{bmatrix} 3 & 1 \\ 1 & 1 \end{bmatrix}$. Then $AB = \begin{bmatrix} 9 & 5 \\ 7 & 5 \end{bmatrix}$. We will pick a matrix C different from this product, say, $C = \begin{bmatrix} 8 & 5 \\ 7 & 5 \end{bmatrix}$. According to the theorem, since $AB \neq C$, $P(ABr = Cr) = P(Dr = 0) \leq 1/2$ where $D = AB - C = \begin{bmatrix} 1 & 0 \\ 0 & 0 \end{bmatrix}$ and r is a vector with entries randomly generated from $\{0, 1\}$. For this simple example, we can compute the exact probability, since the possible vectors r are

$$r = \begin{bmatrix} 0 \\ 0 \end{bmatrix}, \begin{bmatrix} 0 \\ 1 \end{bmatrix}, \begin{bmatrix} 1 \\ 0 \end{bmatrix}, \begin{bmatrix} 1 \\ 1 \end{bmatrix}.$$

Simple matrix vector multiplication shows the products $D\begin{bmatrix} 0 \\ 0 \end{bmatrix}$ and $D\begin{bmatrix} 0 \\ 1 \end{bmatrix}$ are the zero vector, and the others are not. Therefore $P(Dr = 0) = 1/2$. This example shows the upper bound of the inequality in Theorem 1.1 is tight.

1.5   Project 2: Analysis of Freivalds' algorithm

1. By repeating Freivalds' algorithm many times, we can make the probability of failure much smaller. Suppose we apply the algorithm 100 times, by generating random vectors $r^{(1)}, r^{(2)}, \ldots, r^{(100)}$. Discuss when the algorithm will give the wrong answer in this case, and its probability.
2. How would the conclusion of Theorem 1.1 change if the components of the random vector r were chosen uniformly from $\{0, 1, 2\}$? How about from $\{0, 1, \ldots, d\}$?
3. In Theorem 1.1, we proved that $P(Dr = 0) \leq 1/2$ when $D \neq 0$. In Example 1.1, we observed that this bound is tight: the upper bound $1/2$ was attained for a square matrix of size 2. We want to investigate how the maximum value of $P(Dr = 0)$ changes as the dimension of the matrix D increases. In order to do this, we will write a code that estimates $P(Dr = 0)$ where r is a vector matching the dimension of D, with components generated at random uniformly from $\{0, 1, \ldots, d\}$. Here is an outline for a Julia function called **bound** with some hints:

   - The function **bound** should take D, d as inputs. The output will be an approximation for $P(Dr = 0)$. Initialize count $= 0$, and find the dimension of the matrix by $m = \text{size}(D)[1]$.
   - Generate $r^{(1)}, \ldots, r^{(10000)}$ at random, where the components of r are randomly generated numbers from $\{0, 1, \ldots, d\}$. Generating one such vector can be done by the for loop below (notice the code first initializes r as a vector of m 1's, and then overwrites its components):
     ```
     In [ ]: r=ones(m)
                 for j in 1:m
                     r[j]=rand(0:d)
                 end
     ```
   - Count how many vectors $r^{(i)}$ gives $Dr^{(i)} = 0$, where 0 is the zero vector. This can be done by (below **zeros(m)** is a vector of m 0's):
     ```
     In [ ]: if D*r==zeros(m)
                     count=count+1
                 end
     ```
   - Return count/10000.
     Use your code to estimate $P(Dr = 0)$ when D is a randomly generated matrix. Here is how to generate a random matrix of dimension 3, where the matrix entries are randomly generated from $\{0, 1, \ldots, 9\}$:
     ```
     In [1]: D=rand(0:9,3,3)
     Out[1]: 3×3 Array{Int64,2}:
                 7  8  2
                 9  0  6
                 1  4  2
     ```

Experiment with matrices with different dimensions. For each dimension, generate several random matrices to find the maximum value of the various estimates
you get for $P(Dr = 0)$, and check how this maximum changes as the dimension
increases.

1.6 Conditional probability and randomized surveys

The Florida State University football team, the Seminoles, and the University of Florida football team, the Gators, are archrivals. You are hired to find out how many Florida State students are Gator fans in disguise! You cannot just ask plainly, because many students will be embarrassed to tell the truth!

- Can we design a survey with a sensitive question such that the privacy of the respondent is guaranteed?

The answer is yes, but we first need to learn some tools from probability theory.

Definition 1.4 (Conditional probability) The conditional probability that event F occurs given that event E has occurred, denoted by $P(F|E)$, is

$$P(F|E) = \frac{P(F \cap E)}{P(E)} \tag{1.4}$$

provided $P(E) > 0$.

If events F and E are independent, we obtain the following intuitive result:

$$P(F|E) = \frac{P(F \cap E)}{P(E)} = \frac{P(F)P(E)}{P(E)} = P(F),$$

in other words, occurrence of event E has no effect on the probability of event F when events are independent.

Example 1.2 Roll a die, and let E be the event that the outcome is an even number, and F be the event that the outcome is 6. Find $P(F)$, $P(F|E)$, and $P(E|F)$.

Solution 1.1 Clearly, $P(F) = 1/6$. For the other probabilities, we will use the definition of conditional probability:

$$P(F|E) = \frac{P(F \cap E)}{P(E)} \text{ and } P(E|F) = \frac{P(E \cap F)}{P(F)}.$$

The event $F \cap E = E \cap F$ is simply the outcome $\{6\}$, therefore the probability in the numerator is $1/6$. Then we obtain

$$P(F|E) = \frac{1/6}{1/2} = \frac{1}{3} \text{ and } P(E|F) = \frac{1/6}{1/6} = 1.$$

Theorem 1.2 (Law of total probability) *Let $E_1, \ldots, E_n$ be mutually disjoint events in Ω with $\cup_{i=1}^{n} E_i = \Omega$. Then*

$$P(B) = \sum_{i=1}^{n} P(B \cap E_i) = \sum_{i=1}^{n} P(B|E_i)P(E_i).$$

Proof Notice that $B = \cup_{i=1}^{n}(B \cap E_i)$, and since E_i are disjoint, so are $B \cap E_i$, and thus $P(B) = \sum_{i=1}^{n} P(B \cap E_i)$. The second equality follows from the definition of conditional probability. $\qquad\qquad\qquad\qquad\qquad\qquad\qquad\qquad\qquad\qquad\qquad\qquad\qquad\quad$ □

Example 1.3 A Young Scholar decides to skip the math class one morning, and asks his roommate to return his homework to the professor. However, the roommate is not very reliable, and with a 30% chance he will forget to return the homework! Obviously, the Young Scholar gets an F if the homework is not returned. However, this was a tough assignment, and there is 20% chance of failing even if the homework is returned!

1. What is the probability that the Young Scholar will fail the homework?
2. If the Young Scholar fails the homework, what is the probability that it is because the roommate forgot to return it?

Solution 1.2 Let's denote the event that the Young Scholar fails the homework by "Fail", and the event that the roommate forgets to return the homework by "Forget". From the law of total probability,

$$P(\text{Fail}) = P(\text{Fail}|\text{Forget})P(\text{Forget}) + P(\text{Fail}|\text{Not Forget})P(\text{Not Forget})$$
$$= 1 \times 0.3 + 0.2 \times 0.7$$
$$= 0.44.$$

The next probability to compute is

$$P(\text{Forget}|\text{Fail}) = \frac{P(\text{Forget AND Fail})}{P(\text{Fail})}.$$

We know the probability in the denominator from the first part. For the numerator, we use the definition of conditional probability, and rewrite it as

$$P(\text{Forget AND Fail}) = P(\text{Fail}|\text{Forget})P(\text{Forget}) = 0.3.$$

Therefore, $P(\text{Forget}|\text{Fail}) = 0.3/0.44 \approx 0.68$.

Let's go back to our survey design problem. We will discuss two types of surveys designed to handle questions of sensitive nature: the Warner model [25], and the Simmons model [10]. In our example, we want to find the proportion of students who are Gator fans.

1.6.1 Warner model

The survey in this model has two questions answered with a Yes or No.

- **Q1:** I am a Gator fan. Answer Yes or No.
- **Q2:** I am not a Gator fan. Answer Yes or No.

Design of survey: Instruct the student to sit at a table on which there is an ordinary deck of shuffled playing cards (faced down). Student selects one card, privately examines it, and returns it to the deck. Then the deck is reshuffled.

Instructions for the student: If the card drawn is a heart, club, or diamond, answer Q1. If the card is a spade, answer Q2.

In this way, although the interviewer knows the student's response (Yes or No), the interviewer does not know which question the student answered with a Yes, or No. This protects the privacy of the student. Using the law of total probability we have

$$P(yes) = P(yes|Q1 \text{ is chosen })P(Q1 \text{ is chosen}) + P(yes|Q2 \text{ is chosen})P(Q2 \text{ is chosen}).$$
$$(1.5)$$

We observe that

- $P(yes)$—call this number p. This is the proportion of students who answered Yes. We will get this number from the results of the survey.
- $P(yes|Q1 \text{ is chosen})$—this is the proportion we want to know, the proportion of Gator fans; denote this number by g.
- $P(Q1 \text{ is chosen}) = 3/4$.
- $P(yes|Q2 \text{ is chosen})$—this is the proportion of non-Gator fans, which is $1 - g$.
- $P(Q2 \text{ is chosen}) = 1/4$.

From Equation (1.5), we obtain

$$p = \frac{3}{4}g + \frac{1}{4}(1 - g) \Rightarrow g = \frac{4p - 1}{2}.$$

Example 1.4 You survey 100 students and find that 30 students answered Yes, 70 answered No. Then the proportion of students who answered Yes is $p = \frac{30}{100} = 0.3$. From the above formula,

$$g = \frac{4(0.3) - 1}{2} = 0.1.$$

So we estimate the Gator fans to be 10%.

1.6.2 Simmons Model

Some people may still feel uncomfortable with the two questions of the Warner model, since both questions have a sensitive nature. The Simmons model gives the subject a greater sense of security.

In the Simmons model, there are two questions. One is sensitive, but the other one is not. And we know an approximate answer to the nonsensitive one. Consider the following example: Our objective is to find the proportion of high schoolers who have misrepresented their identities in social media.

Design of survey: 1000 high school students were selected at random for the survey. Each student was presented with an unbiased die.

Instructions: Roll the die. If the outcome is 1,2,3, or 4, respond to the question:

- **Q1:** Did you misrepresent your identity in social media during the past 12 months?

If the outcome is 5 or 6, respond to the question:

- **Q2:** Were you born in the month of September?

From the law of total probability, we have

$$P(yes) = P(yes|Q1 \text{ is chosen })P(Q1 \text{ is chosen}) + P(yes|Q2 \text{ is chosen})P(Q2 \text{ is chosen}).$$
(1.6)

We observe

- $P(yes)$—call this p
- $P(yes|Q1 \text{ is chosen})$—this is the proportion we want to know; call it g
- $P(Q1 \text{ is chosen}) = 2/3$.
- $P(yes|Q2 \text{ is chosen})$—yet unknown
- $P(Q2 \text{ is chosen}) = 1/3$

There are two ways to determine $P(yes|Q2 \text{ is chosen})$:

1. Assume a person is equally likely to be born in any month of the year. Then the probability is $\frac{1}{12} \approx 0.083$.
2. Check the birth records of all the students in the high school, if possible, and compute the proportion exactly.

Let's use the first approach so $P(yes|Q2 \text{ is chosen}) = 1/12$. Then, Equation (1.6) simplifies as

$$p = \frac{2}{3}g + \left(\frac{1}{12}\right)\left(\frac{1}{3}\right) \Rightarrow g = \frac{3p}{2} - \frac{1}{24}.$$

Let's assume we conduct this survey and find 80 of 1000 students responded "Yes". Then from the above equation we get

$$g = \frac{3 \times (80/1000)}{2} - \frac{1}{24} \approx 0.078.$$

Therefore, the proportion of students misrepresenting their identity in the past 12 months is approximately 8%.

1.7 Project 3: A survey with three choices

Here is another randomized response model that has three questions. Consider the statements:

$$Q1 : \text{The sensitive statement}$$
$$Q2 : \text{``Yes''}$$
$$Q3 : \text{``No''}$$

Setup: Roll a die. If the outcome is $\{1, 2, 3, 4\}$, answer the sensitive question with a Yes, or No. If the outcome is 5, answer Yes, if the outcome is 6, answer No.

1. Using the law of total probability, derive an equation for the probability of "Yes" for this model.
2. Split into groups. Design a survey using this model. Survey the other group and estimate the proportion you are interested in.

1.8 Bayes' theorem

We start with the conditional probability definition from the previous section:

$$P(F|E) = \frac{P(F \cap E)}{P(E)}. \tag{1.7}$$

Note that

$$P(F \cap E) = P(E|F)P(F),$$

which follows from the definition of $P(E|F)$. Using the law of total probability, Theorem (1.2), where the mutually disjoint events are F and F^c, we write

$$P(E) = P(E|F)P(F) + P(E|F^c)P(F^c)$$

where F^c is the complement of F. Substituting these expressions for $P(F \cap E)$ and $P(E)$ in (1.7) gives

$$P(F|E) = \frac{P(F \cap E)}{P(E)} = \frac{P(E|F)P(F)}{P(E|F)P(F) + P(E|F^c)P(F^c)} \tag{1.8}$$

which is Bayes' theorem (or Bayes' formula).

We can generalize this formula slightly by using n events $F_1, \ldots, F_n$ as the mutually disjoint events with $\cup_{i=1}^{n} F_i = \Omega$ in the law of total probability, instead of F, F^c. Then the formula becomes

$$P(F_1|E) = \frac{P(E|F_1)P(F_1)}{P(E|F_1)P(F_1) + P(E|F_2)P(F_2) + \ldots + P(E|F_n)P(F_n)}.$$

Bayes' theorem (1.8) appeared in a paper by the Reverend Thomas Bayes, published in 1763, and titled *"An Essay towards solving a Problem in the Doctrine of Chances"*. The modern mathematical notation and terminology is so different today that it is not very easy to understand the old texts. For example, Bayes [3] defined probability as follows:

> The probability of any event is the ratio between the value at which an expectation depending on the happening of the event ought to be computed, and the chance of the thing expected upon it's happening.

Example 1.5 Bayes' formula is the principal tool in understanding the probabilities associated with medical tests. Consider a patient taking a medical test to diagnose whether he has a certain sickness. The test could give a positive or negative result, and the patient is either sick or not. This gives four possibilities, described in the next table.

		Actual diagnosis	
		Sick	Healthy
Diagnosis from test	Positive	valid pos	false pos
	Negative	false neg	valid neg

The entry "false positive" means a healthy patient takes the test and the test result is positive. "False negative" means when a sick patient takes the test and the result is negative. The other two entries correspond to situations where the test is accurate.

Assume a particular test has a rate of 99% valid negative and valid positive, for a rare illness only 1% of the population has. If a patient takes this test and the result is positive, what is the probability that the patient is *really* sick?

It is widely reported in the literature that most people give incorrect answers to such questions; see, for example, [14]. The commonly given incorrect answer for this question would be 99%. Once we write the events in question carefully, the error would be obvious.

The question asks for the probability $P(\text{sick}|\text{pos})$. The 99% corresponds to the probability of other events: $P(\text{pos}|\text{sick}) = P(\text{neg}|\text{healthy}) = 0.99$.

To find $P(\text{sick}|\text{pos})$ we use Bayes' formula:

$$
\begin{aligned}
P(\text{sick}|\text{pos}) &= \frac{P(\text{pos}|\text{sick})P(\text{sick})}{P(\text{pos}|\text{sick})P(\text{sick}) + P(\text{pos}|\text{healthy})P(\text{healthy})} \\
&= \frac{0.99 \times 0.01}{0.99 \times 0.01 + 0.01 \times 0.99} \\
&= 0.5.
\end{aligned}
$$

Example 1.6 Bayes' formula is used in computerized testing methods of student exams and essays. Welch and Frick [26] describe several computerized methods that decide whether a student has mastered some learning objective. These methods take several inputs, including the number of correct and incorrect answers the student has given, and the probability that a student who has mastered the learning objective gives a wrong answer to a randomly picked question from the test bank. Let M denote the event that the student has mastered the learning objective, and N be its complement, that is, the student has not mastered it. Let C be the event that the student answers the question correctly. In one example, [26] reports the following values for the probabilities that a student correctly answers a question, when the student has mastered and not mastered the topic:

$$P(C|M) = 0.85, \quad P(C|N) = 0.40.$$

Note that these probabilities imply the following probabilities for the complementary events:

$$P(C^c|M) = 0.15, \quad P(C^c|N) = 0.60$$

where C^c is the complement of C, meaning the student has answered the question incorrectly.

Let's assume that the probability that a randomly picked student has mastered the learning objective is 0.5, that is, $P(M) = P(N) = 0.5$. These probabilities are called *prior* probabilities; it is our best guess in the absence of any information. Now assume that the student is presented a randomly selected question, and the student answers it *incorrectly*. What is the probability that the student actually has mastered the topic?

The probability we want to know is $P(M|C^c)$. From Bayes' formula

$$P(M|C^c) = \frac{P(C^c|M)P(M)}{P(C^c|M)P(M) + P(C^c|N)P(N)}$$

$$= \frac{0.15 \times 0.5}{0.15 \times 0.5 + 0.6 \times 0.5}$$

$$= 0.2.$$

Recall that our initial estimate for $P(M)$ was 0.5, and now with the student answering the first question incorrectly, our updated probability that the student has mastered the topic went down to 0.2.

Now let's assume the student gets another question, testing the same learning objective, and answers it correctly this time. How will the probability that the student has mastered the topic change? We want to compute

$$P(M|C) = \frac{P(C|M)P(M)}{P(C|M)P(M) + P(C|N)P(N)},$$

but now, we will not assume $P(M) = P(N) = 0.5$ as before. We have updated information about the student who did not answer the first question correctly. The *posterior* probability obtained in the previous step, $P(M|C^c) = 0.2$, will be used as the new value for $P(M)$, and similarly, $P(N|C^c) = 0.8$ is the new value for $P(N)$, in the second application of Bayes' formula:

$$P(M|C) = \frac{P(C|M)P(M)}{P(C|M)P(M) + P(C|N)P(N)}$$

$$= \frac{0.85 \times 0.2}{0.85 \times 0.2 + 0.4 \times 0.8}$$

$$= 0.35.$$

The probability that the student has mastered the topic has increased from 0.20 to 0.35 after answering the second question correctly.

1.9 Project 4: The Haunting of Hill House

The Haunting of Hill House [11] is a
scary book about a haunted house written
by Shirley Jackson. There are ghosts in
this house, and they bang on the walls and
the doors at night, open doors that were
shut closed, and do many other scary
things. A group of people visit the house
to uncover its secrets, among which is
Mrs. Montague, who is not afraid the
least, and on the contrary, on a quest to
find the ghosts and communicate with
them. One other eerie detail about this
house is that it has a kitchen with three
doors in it.

We will now deviate from the original story a little, and assume there are two
ghosts, one good and one bad, and one night as Mrs. Montague is drinking hot
chocolate in the kitchen, the bad ghost starts banging on the doors. Mrs. Montague
knows the ghost is behind one of the doors, even though the banging seems to be
coming from all around, and if she can just open the right door she might actually
catch the ghost before it can flee. She thinks to herself that her chances for finding
the right door is 1/3.

But now our story gets more complex, and the good ghost, who wants Mrs.
Montague to find the bad ghost, enters the scene. The good ghost cannot simply
open the door the bad ghost is hiding behind, for she does not want to incur the
bad ghost's wrath. She decides to help Mrs. Montague in the following way: once
Mrs. Montague approaches one of the doors to open it, the good ghost opens one
of the remaining two doors that do not hide the bad ghost. She hopes that this extra
information will help Mrs. Montague to find the door that hides the bad ghost.

You need to help Mrs. Montague, if you are not too scared, with her decision.
Let's try to simplify the problem a little. Once she makes her choice, Mrs. Montague
will approach one of the doors to open it—let's call that Door 1. Then the good
ghost will open one of the remaining doors (that does not reveal the bad ghost). Mrs.
Montague will see that there is no ghost behind the door the good ghost opened,
and at that point she will have two options: open Door 1 as she intended in the first
place, or change her decision and open the door the good ghost did not open. So the
question is whether Mrs. Montague should stick with her original choice, or switch
to the other door? Is the information provided by the good ghost really helpful or
not?

Let G_1 be the event that the ghost is behind Door 1, the door Mrs. Montague is
inclined to open. Let's label the other doors as Door 2 and Door 3 (it is not important
which one is labeled as 2 or 3). Since we assume the bad ghost is equally likely
to be hiding behind any of the three doors, $P(G_1) = P(G_2) = P(G_3) = 1/3$. Then

the good ghost opens a door; this will be either Door 2 or Door 3. Assume the good ghost opened Door 2. Then we need to compute $P(G_1 \mid$ Door 2 opened), the probability that the bad ghost is behind Door 1, after Mrs. Montague receives the new information. Notice that after Door 2 is opened, there are two doors the bad ghost might be hiding behind: Door 1, or Door 3. If $P(G_1 \mid$ Door 2 opened) and $P(G_3 \mid$ Door 2 opened) are 1/2, then Mrs. Montague does not need to switch to the other door. If $P(G_3 \mid$ Door 2 opened) is larger than 1/2, then Mrs. Montague should switch to Door 3. A similar reasoning applies if the door the good ghost opened was Door 3.

1. Play out the situation described by this problem, or write a Julia code to simulate it, to estimate the probability of finding the bad ghost when Mrs. Montague switches the door she initially picked.
2. Use Bayes' theorem to compute $P(G_1 \mid$Door 2 opened) and $P(G_3 \mid$Door 2 opened). Should Mrs. Montague switch the door or not?
3. Let us modify the problem and assume the good ghost is not afraid of the bad one, and simply opens either Door 2 or Door 3 at random with equal probabilities. Compute $P(G_1 \mid$Door 2 opened), $P(G_2 \mid$Door 2 opened), and $P(G_3 \mid$Door 2 opened). Should Mrs. Montague switch the door or not in this situation?

Chapter 2
Discrete Random Variables

In many problems involving probability, we are not as much interested in the actual random outcome, but a *function* of this outcome. For example, in Section 1.4 where we discussed Freivalds' algorithm, we considered a sample space of vectors $r = [r_1, \ldots, r_n]^T$ where each r_i was either 0 or 1. We then picked a vector r at random from the sample space, however, our main interest was not the actual outcome r but Dr, and specifically whether Dr is the zero vector, where D was a square matrix of size n. Figure 2.1 describes this schematically: X is a function from the sample space of all vectors r with components 0 or 1, to the space $\mathbb{R}^n$, defined by $X(r) = Dr$.

In probability theory we call the function X a **multivariate random variable**. If the range is the real numbers $\mathbb{R}$, that is $n = 1$, we say X is a **random variable**. In this book, we will mostly consider the one-dimensional case $n = 1$. Note that the range of X in the above example is a finite set: it consists of all vectors $Dr^{(1)}, Dr^{(2)}, \ldots$, and there are finitely many of them. When the values X can attain are finite, or countable, we say the random variable is **discrete**. If the values X can attain are uncountable, we say it is a **continuous** random variable. We will study discrete random variables in this chapter, and continuous random variables in the next one.

Definition 2.1 (Discrete random variable) A random variable on a sample space Ω is a real-valued function X on Ω, that is, $X : \Omega \to \mathbb{R}$. If X takes on countably many values, it's called a discrete random variable.

Example 2.1 Let Ω be the outcomes when two dice are rolled, $\{(1, 1), (1, 2), (2, 1), \ldots\}$, and let X be the sum of the numbers on the dice. Then X is a discrete random variable.

We write $\{X = a\}$ for the event that consists of all outcomes for which X takes the value a. In the above example, $\{X = 4\}$ corresponds to the event $\{(1, 3), (3, 1), (2, 2)\}$. What is quite interesting is now we can assign a probability to $\{X = 4\}$: it will be equal to the probability of the event $\{(1, 3), (3, 1), (2, 2)\}$:

$$P\{X = 4\} = P(\{(1, 3), (3, 1), (2, 2)\}) = 3/36.$$

© The Editor(s) (if applicable) and The Author(s), under exclusive
license to Springer Nature Switzerland AG 2020
G. Ökten, *Probability and Simulation*, Springer Undergraduate Texts
in Mathematics and Technology, https://doi.org/10.1007/978-3-030-56070-6_2

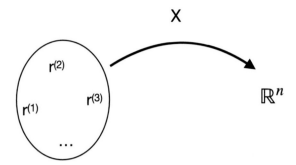

Domain of X: Sample space of outcomes   **Range of X**: A subset of $\mathbb{R}^n$

Fig. 2.1: The definition of random variable

What about $P\{X = 1\}$? The smallest the sum of two dice can be is 2, so this event is not possible. Therefore we set $P\{X = 1\} = 0$.

Definition 2.2 (Support) The set of real numbers x such that $P\{X = x\} > 0$ is called the support of the random variable, and denoted by $\mathcal{S}$.

Most questions involving a discrete random variable can be answered using its **probability mass function**.

Definition 2.3 (Probability mass function) The probability mass function (pmf) of a discrete random variable X is defined as

$$f(x) = P\{X = x\}.$$

The mass function satisfies

1. $f(x) > 0$ if $x \in \mathcal{S}$ (if $x \notin \mathcal{S}$, then $f(x)$ is set to 0),
2. $\sum_{x \in \mathcal{S}} f(x) = 1$,
3. $P(X \in A) = \sum_{x \in A} f(x)$ where $A \subseteq \mathcal{S}$.

We learned about independent events in Chapter 1. This concept generalizes to random variables.

Definition 2.4 (Independent random variables) Random variables X and Y are independent if
$$P\{X = x, Y = y\} = P\{X = x\}P\{Y = y\}$$

for all $x, y \in \mathbb{R}$. We define similarly mutual independence of $X_1, \ldots, X_k$.

Definition 2.5 (Expectation) The expected value (also called the expectation, or the mean) of a discrete random variable X, denoted by $E[X]$, is

$$E[X] = \sum_{i \in S} iP\{X = i\} = \sum_{i \in S} if(i)$$

where the summation is over all values of X, and f is the probability mass function of X.

The expected value is like an average. Suppose X took on the values $1, 2, 3, 4$ with equal probabilities $1/4$. Then the above formula gives

$$E[X] = 1 \times (1/4) + 2 \times (1/4) + 3 \times (1/4) + 4 \times (1/4) = (1 + 2 + 3 + 4)/4$$

which is precisely the average of $1, 2, 3, 4$. If the probabilities are not equal, then the expectation can be thought as a "weighted" average, where each value is multiplied by its weight, that is, its probability.

Exercise 2.1 Solve the following problems.

1. Find the expected value of X where X is the outcome of a die.
2. What is the expected value of a constant?

Note 2.1 (Expectation is linear) The expectation operation satisfies the following properties:

1. $E[\sum_{i=1}^{n} X_i] = \sum_{i=1}^{n} E[X_i]$.
2. For any $c \in \mathbb{R}$, $E[cX] = cE[X]$.

2.0.1 Expectation of a function of a random variable

Suppose that we are given a discrete random variable X and its probability mass function $f(x)$. We want to compute the expected value of a *function of* X, say $g(X)$. First we observe that $g(X)$, which is the composition of two functions g and X, is itself a random variable. Therefore, we can think about the expected value of $g(X)$ in the same way we did before: it is the sum of the function values multiplied by the corresponding probabilities.

Example 2.2 Let X be a random variable that takes values $-1, 0, 1$, with probabilities $0.2, 0.5, 0.3$, respectively. Compute $E[X^2]$.
Let $Y = X^2$. Observe that Y takes on values $(-1)^2, 0^2, 1^2$, that is, 0 and 1. The probabilities are

$$P\{Y = 0\} = P\{X = 0\} = 0.5$$
$$P\{Y = 1\} = P\{X = -1\} + P\{X = 1\} = 0.5.$$

Therefore,

$$E[X^2] = E[Y] = 0(0.5) + 1(0.5) = 0.5.$$

Here is the formal definition of the expected value of a function of a random variable.

Definition 2.6 If $f(x)$ is the probability mass function of a discrete random variable X with support S, then for any real-valued function g

$$E[g(X)] = \sum_{x \in S} g(x) f(x).$$

Definition 2.7 (Variance) The variance of a random variable X is

$$Var(X) = E[(X - E[X])^2] = E[X^2] - (E[X])^2.$$

Another common notation for variance is σ^2. The square root of variance, σ, is called the **standard deviation**. Variance measures how much the random variable X deviates from its expected value.

For example, the random variables X, Y

$$X = \begin{cases} -1, & \text{with probability } 1/2 \\ 1, & \text{with probability } 1/2 \end{cases}, \quad Y = \begin{cases} -10, & \text{with probability } 1/2 \\ 10, & \text{with probability } 1/2 \end{cases}$$

have the same expected value, 0, but Y deviates much more from its mean than X:

$$Var(X) = E[X^2] - (E[X])^2 = E[X^2] = 1\frac{1}{2} + 1\frac{1}{2} = 1$$

$$Var(Y) = E[Y^2] - (E[Y])^2 = E[Y^2] = 100\frac{1}{2} + 100\frac{1}{2} = 100.$$

Exercise 2.2 Show that $Var(aX + b) = a^2 Var(X)$.

Theorem 2.1 *If X and Y are independent, then*

1. $E[XY] = E[X]E[Y]$,
2. $Var(X + Y) = VarX + VarY$.

Proof The first statement follows from the definitions of expectation and independent random variables. Let S_1, S_2 be the support of X and Y. Then

$$E[XY] = \sum_{i \in S_1} \sum_{j \in S_2} ij P\{X = i, Y = j\} = \sum_{i \in S_1} \sum_{j \in S_2} ij P\{X = i\} P\{Y = j\}$$

$$= \left(\sum_{i \in S_1} i P\{X = i\} \right) \left(\sum_{j \in S_2} j P\{Y = j\} \right)$$

$$= E[X]E[Y].$$

For the second statement, use the definition of variance to write

$$Var(X + Y) = E[(X + Y)^2] - (E[X] + E[Y])^2$$
$$= E[X^2] + E[Y^2] + 2E[XY] - E[X]^2 - E[Y]^2 - 2E[X]E[Y].$$

Rearranging the right-hand side we obtain

$$(E[X^2] - E[X]^2) + (E[Y^2] - E[Y]^2) + 2(E[XY] - E[X]E[Y]).$$

From the first part of the theorem, $2(E[XY] - E[X]E[Y]) = 0$, and the remaining terms are equal to $VarX + VarY$. Therefore, we have proved

$$Var(X + Y) = VarX + VarY$$

when the random variables are independent.   □

2.1 Discrete uniform random variables

Let X be a random variable that takes values $1, 2, \ldots, n$. We say X is a **discrete uniform random variable** (or, X has a **discrete uniform distribution**) on the integers $1, 2, \ldots, n$ if $P\{X = i\} = \frac{1}{n}$ for all $i = 1, 2, \ldots, n$.

Lemma 2.1 *If X is a discrete uniform random variable on $1, 2, \ldots, n$ then $E[X] = \frac{n+1}{2}$.*

Proof Simply observe $E[X] = \frac{1}{n}(1 + \ldots + n) = \frac{1}{n}\frac{n(n+1)}{2} = \frac{n+1}{2}$.   □

Example 2.3 (Random Harmonic Series)
    In Calculus, you learned that the harmonic series $\sum_{n=1}^{\infty} \frac{1}{n}$ is divergent. On the other hand, the alternating series $\sum_{n=1}^{\infty} \frac{(-1)^{n-1}}{n}$ is convergent:

$$1 + \frac{1}{2} + \frac{1}{3} + \ldots = \infty$$

$$1 - \frac{1}{2} + \frac{1}{3} - \ldots = \log 2$$

Consider the following generalization: $\sum_{n=1}^{\infty} \frac{a_n}{n}$ where a_n is equal to 1 or -1 with probability $1/2$ (in other words, a_n is a discrete uniform random variable on $-1, 1$.) What can be said about the convergence of this **random harmonic series**?
    Let's investigate this question numerically. Using Julia, we will generate 50000 random a_n's to compute $S = \sum_{n=1}^{50000} \frac{a_n}{n}$. We will then repeat this 1000 times to obtain $S^{(1)}, S^{(2)}, \ldots, S^{(1000)}$, and plot a histogram for these values.

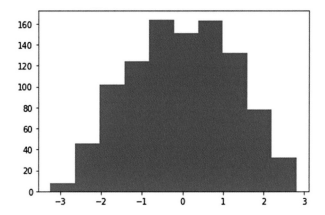

In [1]: using PyPlot

We first write a function, **rnd**, which generates the numbers −1 and 1 at random with equal probability:

```
In [2]: function rnd()
            x=rand()
            if x <0.5
                -1.0
            else
                1.0
            end
        end
```

Out[2]: rnd (generic function with 1 method)

The next function **psum** computes the partial sum $S = \sum_{k=1}^{50000} \frac{a_k}{k}$:

```
In [3]: function psum(n)
            sums=Array{Float64}(undef,0)
            i=1
            while i<=n
            s=0.0
                for k in 1:50000
                    s=s+rnd()/k
                end
            append!(sums,s)
            i=i+1
            end
            hist(sums,10)
        end
```

Out[3]: psum (generic function with 1 method)

Now we compute 1000 partial sums and plot their histogram.

In [4]: psum(1000);

According to the histogram, all the partial sums seem to be between -3 and 3. This numerical evidence may be in support of the convergence of the random harmonic series, however, the definitive answer can be given only by theoretical means. In fact, Hans Rademacher[1] proved the convergence of this series in 1922.

Theorem 2.2 (Rademacher, 1922) *If $\sum_{n=1}^{\infty} c_n^2 < \infty$, then $\sum_{n=1}^{\infty} a_n c_n$ exists with probability 1, where $a_n = 1$ or -1 with probability 1/2.*

---

[1] Rademacher, H., 1922. Einige Sätze über Reihen von allgemeinen Orthogonalfunktionen. Mathematische Annalen, 87(1-2), pp. 112–138.

2.2 Project 5: Benford's law

Data is money, it is precious. It runs the engines of many businesses in today's economy. Here is a riddle involving data: if you wrote down the population of all cities, or counties, in the U.S., and examined the first digits of these numbers, would you see each number 1 through 9 appearing as the first digit in roughly equal proportions, or would some numbers appear more often than others? To investigate this question, we download the county population data from the U.S. Census Bureau web page (http://factfinder.census.gov), and use Julia to analyze the distribution of the digits. The data "population_county.csv" and the Julia notebook "Benford's Law US population.ipynb" are available on the Springer web page for the book. The Julia notebook can also be found in Appendix A. Here we summarize the results obtained in the Julia notebook.

A simple graphical approach to get a sense of the digit distribution is to plot its histogram. There are 3142 numbers in our data. We extract the first digits of these numbers, and plot their relative frequency histogram. We repeat this experiment for the second and third digits as well. (A few populations with two digits were removed from the data.) Figure 2.2 plots these histograms. The top-left histogram is for the first digits, and it clearly shows the distribution is not uniform. Smaller digits appear more often than the larger ones. The distribution of the second digits (top right) seems to have the same pattern, but less pronounced. The pattern seems to disappear by the third digits: the histogram for the third digits (bottom left) and the histogram of 3142 random integers 0 through 9 from the uniform distribution (bottom right) look very alike.

Similar observations were made by Newcomb [17] in 1881, and Benford [4] in 1938, when they examined the digit distributions of some data sets. Newcomb suggested the following probability mass function for the first digit

$$P(\text{first digit } = d) = \log_{10}\left(1 + \frac{1}{d}\right) \tag{2.1}$$

for $d = 1, 2, \ldots, 9$. Benford gave examples of many data sets that follow this probability mass function; this phenomenon is known as the Benford's law, or Benford–Newcomb law.

Newcomb also investigated the distribution of the second leading digits. He suggested the following probabilities for them

$$P(\text{second digit } = d) = \sum_{k=1}^{9} \log_{10}\left(1 + \frac{1}{10k + d}\right)$$

for $d = 0, 1, \ldots, 9$.

When do we expect data to follow Benford's law? Some intuitive explanations are

- When data describe the sizes of similar phenomena, with no artificial minimum or maximum constraints, such as weights, distances, temperatures, and dollars,

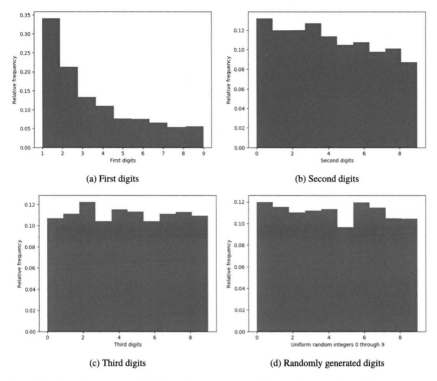

(a) First digits

(b) Second digits

(c) Third digits

(d) Randomly generated digits

Fig. 2.2: Distributions of digits of county population data and uniform random numbers

of things. For such data, units are irrelevant, that is data should follow the law independent of the measurement units.

- When data come from different distributions, that is data is a mixture of numbers coming from different populations with different distributions.

Figure 2.3 plots the relative frequency histogram for the first digits of the population data together with the probability mass function (2.1) for the first digits. The fit of the mass function to the data is quite well.

Benford's law can be used to detect fraud. For example, it is widely reported that accounting data follows Benford's law. Then a significant deviation from Benford's law might point to possible fraud. Applications of Benford's law to forensic accounting is the subject of a book by Nigrini [18]. Some researchers claim Benford's law for the first digits can also be used to detect election fraud, although not everyone agrees. Some argue that Benford's law for the second digits is a better choice for election data. Figure 2.4 plots the county by county votes for Democrats in the 2016 U.S. presidential election together with the Benford's probabilities for the first digit, and the fit to the Benford's law looks perfect.

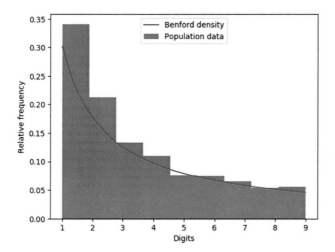

Fig. 2.3: Distribution of first digits of county population data with Benford's probability mass function for the first digits

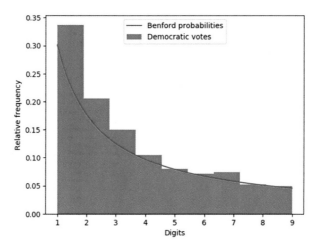

Fig. 2.4: Distribution of first digits of county by county votes for Democrats in 2016 election

Iran's 2009 presidential election was very controversial. The incumbent Mr. Ahmadinejad won the election with 62% of the vote, while his main challenger Mr. Mousavi received 34% of the vote. Critics claimed widespread election fraud, and street protests started shortly after the election results were announced.

The data "Iran_pres_2009.csv" contains the votes from the ballot boxes, and was obtained from http://thomaslotze.com/iran/#Sevens. Download the data and the Julia notebook "Benford's Law US population.ipynb", from the Springer web page for the book, and save them into the same directory. If you never used JuliaDB before, download the package by first executing **Pkg.add("JuliaDB")**, and then running **using JuliaDB**. Modify the Julia notebook "Benford's Law US population.ipynb" to answer the following questions:

1. The number of votes received by Mr. Ahmadinejad and Mr. Mousavi are given in the columns with headings **Ahmadinejad** and **Mousavi**. Plot relative frequency histograms for the first digits of the votes. Then superimpose the Benford probability mass function for the first digits to the histograms. (In some ballot boxes candidates received no votes. To remove the zeros from the first digit data, use **first_digs_nz=filter(x->x>0,first_digs)**.)
2. Comment on the fit of the data to Benford's law for Mr. Ahmadinejad and Mr. Mousavi. Do you think the results support possible election fraud?

2.3 Bernoulli, binomial, geometric, Poisson random variables

In this section, we will learn about some discrete random variables that frequently appear in applications. When we encounter a new random variable, we should think about the following questions: what are the values the random variable can take, and what are the corresponding probabilities? This information is encapsulated in the probability mass function of the random variable.

2.3.1 Bernoulli and binomial random variables

The Bernoulli and binomial random variables were introduced by Jacob Bernoulli in his book *Ars Conjectandi* (The Art of Conjecturing), published in 1713. Consider a random experiment with only two outcomes. One outcome is labeled as "success" and has probability p, the other is labeled as "failure" and has probability $1 - p$. Let $Y = 1$ if the outcome is a success and $Y = 0$ if it is a failure. Then Y is called a **Bernoulli random variable** with success probability p, and denoted by $Y \sim Ber(p)$. Its probability mass function is

$$f(0) = P\{Y = 0\} = 1 - p \text{ and } f(1) = P\{Y = 1\} = p.$$

Computing the expectation and variance of $Y \sim Ber(p)$ is simple:

$$E[Y] = 0(1 - p) + 1p = p$$

$$Var(Y) = E[Y^2] - E[Y]^2 = p - p^2 = p(1 - p).$$

Now consider repeating this random experiment independently, n times, and label the resulting Bernoulli random variables (also called Bernoulli trials) as $Y_1, \ldots, Y_n$. We are interested in the number of successes obtained; call this value X. Note that X can be written as the sum of the Y_i's:

$$X = \text{ number of successes in } n \text{ trials} = \sum_{i=1}^{n} Y_i$$

where $Y_i = 1$ if the ith trial is success and 0 otherwise. Then X is called a **binomial random variable** with parameters n, p and denoted by $X \sim Bin(n, p)$.

Observe that X takes on values $0, 1, \ldots, n$, with the probability mass function

$$f(i) = P\{X = i\} = \binom{n}{i} p^i (1 - p)^{n-i}, \tag{2.2}$$

where $\binom{n}{i}$ is the binomial coefficient, and it counts the number of different ways one can have i successes in a sequence of n trials. Since each arrangement of i successes in n trials (thus $n - i$ failures) has the probability $p^i (1 - p)^{n-i}$, adding these probabilities $\binom{n}{i}$ times leads to Eq. (2.2).

Lemma 2.2 *The expected value of $X \sim Bin(n, p)$ is $E[X] = np$. Its variance is $Var(X) = np(1 - p)$.*

Proof Since $X = \sum_{i=1}^{n} Y_i$ and the Y_i's are independent Bernoulli random variables with success probability p, we have

$$E[X] = \sum_{i=1}^{n} E[Y_i] = np$$

and

$$Var(X) = \sum_{i=1}^{n} Var(Y_i) = np(1 - p).$$

2.3.2 Geometric random variables

Suppose that independent Bernoulli trials, each having a probability of success p, are performed until a success occurs. Let X be the number of trials required until the first success. Then we call X a **geometric random variable** with parameter p, and write $X \sim Geo(p)$. Observe that X takes on values $1, 2, \ldots$, with the probability mass function

$$f(i) = P\{X = i\} = (1 - p)^{i-1}p$$

for $i = 1, 2, \ldots$. To see why $P\{X = i\} = (1 - p)^{i-1}p$, simply observe that the event $\{X = i\}$ means the first $i - 1$ trials are failures, which has probability $(1 - p)^{i-1}$, and the ith trial is a success, which has probability p.

Exercise 2.3 Show that the above probability mass function satisfies the properties of a mass function.

Lemma 2.3 *The expected value of $X \sim Geo(p)$ is $\frac{1}{p}$. Its variance is $Var(X) = \frac{1-p}{p^2}$.*

Proof Let's prove the first part of the lemma. We have

$$E[X] = \sum_{i=1}^{\infty} i(1 - p)^{i-1}p = p \sum_{i=1}^{\infty} iq^{i-1}, \qquad (2.3)$$

where $q = 1 - p$. Recall the geometric series from calculus

$$1 + q + q^2 + \ldots = \sum_{i=0}^{\infty} q^i = \frac{1}{1 - q}.$$

We will differentiate each expression in the above equation with respect to q. Recall from calculus that the geometric series can be differentiated term by term if $|q| < 1$. We obtain

$$1 + 2q + 3q^2 + \ldots = \sum_{i=1}^{\infty} iq^{i-1} = \frac{d}{dq}\left(\frac{1}{1-q}\right) = (1-q)^{-2} = \frac{1}{p^2}.$$

Substituting $\sum_{i=1}^{\infty} iq^{i-1}$ by $1/p^2$ in Eq. (2.3), we get $E[X] = p\frac{1}{p^2} = \frac{1}{p}$.

Lemma 2.4 *If $X \sim Geo(p)$, then $P\{X \geq n\} = (1-p)^{n-1}$.*

Proof The event $\{X \geq n\}$ means the first success occurs either at trial $n, n+1, \ldots$. Or, equivalently, the first success does not occur in the first $n-1$ trials. That's saying the first $n-1$ trials are all failures, which happens with probability $(1-p)^{n-1}$. □

Lemma 2.5 (Memoryless property)   *If X is a geometric random variable with parameter p, and $n, k > 0$, then*

$$P\{X = n + k \mid X > k\} = P\{X = n\}. \tag{2.4}$$

Proof Using the definition of conditional probability, we write

$$P\{X = n + k \mid X > k\} = \frac{P\{X = n + k, X > k\}}{P\{X > k\}}.$$

The event $\{X = n + k, X > k\}$ is the intersection of the events $\{X = n + k\}$ and $\{X > k\} = \{X = k + 1, \ldots, X = k + n - 1, X = k + n, X = k + n + 1, \ldots\}$, and this intersection is $\{X = n + k\}$. Also note that $\{X > k\}$ can be rewritten as $\{X \geq k + 1\}$. Then we get

$$P\{X = n + k \mid X > k\} = \frac{P\{X = n + k, X > k\}}{P\{X > k\}} = \frac{P\{X = n + k\}}{P\{X \geq k + 1\}}.$$

From the definition of geometric random variable, $P\{X = n + k\} = (1-p)^{n+k-1}p$, and from Lemma 2.4, $P\{X \geq k + 1\} = (1-p)^k$. Their ratio is

$$\frac{P\{X = n + k\}}{P\{X \geq k + 1\}} = \frac{(1-p)^{n+k-1}p}{(1-p)^k} = (1-p)^{n-1}p = P\{X = n\},$$

completing the proof.   □

We used a strange expression, "memoryless property", in the statement of Lemma 2.5. What does that mean? Let's say we are flipping a coin repeatedly until we obtain tails, and X counts the number of flips until the first success, that is, the first tail. We know X is a geometric random variable. Now imagine the first k flips are all heads, in other words $\{X > k\}$. We will flip the coin n more times, and want to know the probability that the first success will happen at the $(n + k)$th flip. This is the probability $P\{X = n + k \mid X > k\}$ on the left-hand side of Eq. (2.4). Going back to our coin experiment, to the moment when we are ready to flip the coin the $(k + 1)$st time, does the coin remember that it gave failures the first k flips, and perhaps will have pity on us and bring a success quickly? Of course not! As far as the coin is concerned, we are as if restarting the counting until the first success. Then the probability that the first success will happen at the $(n + k)$th flip, conditional on having all failures the first k flips, must be the same as the probability that the first success will happen at the nth flip when we restart the counting, which is the right-hand side of Eq. (2.4).

2.3.3 Poisson random variables

The Poisson distribution was introduced by S. D. Poisson in a book[2] he wrote in 1837, regarding the applications of the probability theory to lawsuits, criminal trials, and the like. In many experiments, we count the number of times a particular event occurs during a time period. For example, we could count the number of customers arriving at a store between 9AM and 12PM, or the number of spam phone calls received during a week. If certain conditions are satisfied, the number of occurrences can be modeled using the Poisson random variable. We say X is a **Poisson random variable** with parameter $\lambda > 0$ if

$$f(i) = P\{X = i\} = e^{-\lambda}\frac{\lambda^i}{i!}$$

where $i = 0, 1, \ldots$.

Exercise 2.4 Show that the Poisson probability mass function satisfies the properties of a probability mass function.

Lemma 2.6 Let X be a Poisson random variable with parameter λ. Then

$$E[X] = \lambda \text{ and } Var(X) = \lambda.$$

Proof We prove the first statement. Simply observe

$$E[X] = \sum_{i=0}^{\infty} ie^{-\lambda}\frac{\lambda^i}{i!} = e^{-\lambda}\sum_{i=1}^{\infty}i\frac{\lambda\lambda^{i-1}}{i(i-1)!} = \lambda e^{-\lambda}\sum_{i=1}^{\infty}\frac{\lambda^{i-1}}{(i-1)!} = \lambda e^{-\lambda}e^{\lambda} = \lambda.$$

Example 2.4 One of the earliest applications of Poisson distribution to real-world data was by Ladislaus Bortkiewicz, who published a book[3] on probability theory in 1898. In the book he presented data on the number of soldiers in the Prussian army who were kicked to death by horses over a 20-year period (1875–1894). He considered 14 army corps, and counted the number of deaths during each year, in each corp, for 20 years, giving him 280 data points. Table 2.1, taken from [24], summarizes the data:

---

[2] Poisson, S.D., 1837. Recherches sur la probabilite des jugements en matiere criminelle et en matiere civile precedees des regles generales du calcul des probabilites. Paris, France: Bachelier.

[3] von Bortkiewicz, L., 1898. Das Gesetz der kleinen Zahlen (The Law of Small Numbers). BG Teubner, Vancouver.

Number of Deaths	0	1	2	3	4	≥ 5
Observations		144	91	32	11	2	0

Table 2.1: Deaths per year

The entry 32 in the table means over the period of 20 years, 32 army corps had 2 deaths by horse kicks. Bortkiewicz's hypothesis was the number of deaths followed the Poisson distribution very closely. Let's investigate his claim.

Let X be the number of deaths from horse kicks. From the data, we see that X takes on the values 0 through 4 with the specified frequencies, from which we can obtain the relative frequencies. Since the total number of observations is 280, the relative frequency of 2 deaths, for example, is $32/280 \approx 0.11$. We can also approximate the expected value of X from the data by computing the average number of deaths as $\frac{0 \times 144 + 1 \times 91 + 2 \times 32 + 3 \times 11 + 4 \times 2}{280}$ which is 0.7.

If we suspect X follows the Poisson distribution, then what is its parameter λ? Since the expected value of X is λ, and since from the data we computed the average as 0.7, it makes sense to set $\lambda = 0.7$. Now we can compute the probabilities $P(X = n), n = 0, 1, 2, 3, 4$, and compare them with the relative frequencies obtained from the data. For example,

$$P(X = 2) = e^{-0.7} \frac{(0.7)^2}{2!} = 0.122$$

and from the data the relative frequency of 2 deaths is $32/280 = 0.114$. The following table displays the Poisson probabilities and relative frequencies for all the cases. The probabilities and the relative frequencies are pretty close to each other (Table 2.2).

	$n = 0$	$n = 1$	$n = 2$	$n = 3$	$n = 4$	$n \geq 5$
$P(X = n)$	0.497	0.348	0.122	0.0284	0.00497	0.0007
Relative frequency	0.514	0.325	0.114	0.0393	0.00714	0

Table 2.2: Poisson probabilities versus relative frequencies for horse kick data

The horse kick data seems to follow the Poisson distribution closely, but how can we tell in general what kind of data follows the Poisson distribution? Here is an informal recipe: the number of occurrences (in the horse kick data an occurrence is a death) follows a Poisson random variable with some parameter λ if

- The number of occurrences in nonoverlapping intervals are independent;
- The probability of exactly one change in a sufficiently short interval of length h is approximately λh;
- The probability of two or more changes in a sufficiently short interval is essentially zero.

2.4 Project 6: Resurrect the Beetle!

Let's play the game *Resurrect the Beetle!*, a game adapted from Maull and Berry [13] (Fig. 2.5).

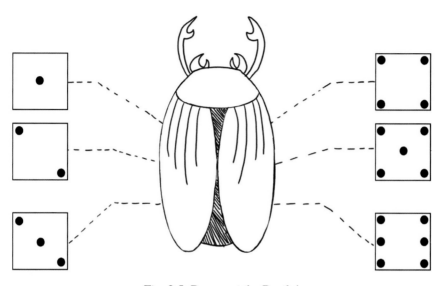

Fig. 2.5: Resurrect the Beetle!

1. We will play the game with 6 players, although any number of players is possible. Players will take turns to roll a die. Say the first player rolls 5. Then the player draws the corresponding leg of the beetle. The second player rolls next. If she rolls 5, then she simply passes the die to the next person, since the leg corresponding to 5 is already drawn. In general, a player draws the leg corresponding to what she rolled, only if that leg is not already drawn. We want to estimate how many rolls are required to draw all the legs, and resurrect the beetle! As you play this game, complete the following table to answer a few more questions.

Number of rolls

1st leg	2nd leg, after one drawn	3rd leg, after two drawn	4th leg, after three drawn	5th leg after four drawn	6th leg after five drawn
1					

2. The coupon collector's problem is a classical problem in probability theory going back to the eighteenth century. Here is a breakfast-friendly description of the problem: imagine the following contest where you buy cereal boxes, and

every cereal box contains a coupon. There are n types of coupons. The coupon in each cereal box is chosen at random uniformly from these n possibilities. Let's label the coupons as $c_1, c_2, \ldots, c_n$. How many boxes of cereal must one buy before obtaining at least one coupon of each type?

Let X denote the number of boxes one has to buy before collecting one coupon from each type. Clearly, X is a random variable. We want to find $E[X]$.

Hint: Let X_1 be the number of boxes needed to get the first coupon, X_2 be the number of boxes needed to get the second distinct coupon, when we have already one coupon, X_3 be the number of boxes needed to get the third distinct coupon, when we have already two types of coupons, etc. What can you say about the distribution of $X_1, X_2, X_3, \ldots, X_n$? How do you think X is related to these random variables?

3. Can you think of the beetle game in terms of the coupon collector's problem? What is the theoretical number of rolls required to draw all the legs based on your analysis in part (2)?

4. The song repertoire of a bird is thought to indicate its overall health, age, and fitness. Researchers try to find the size of the song repertoire of a bird by recording its songs, and analyzing song elements using spectrograms. For example, Balsby and Hansen [2] examine about 5500 song elements of a 2-year-old male whitethroat, and find that 625 of them are distinct song elements. But is this the limit of the song repertoire of the bird? If the authors had more recordings, could they have found more distinct song elements?

   a. Discuss how you could use the coupon collector's problem as a model for the song repertoire problem, and explain the assumptions you make to that end. The quantities of interest are the number of distinct song elements a bird knows, and the number of observations needed to observe all the distinct song elements.

   b. Assume the actual number of distinct song elements is indeed 625 as the authors claim. Using the coupon collector's problem as the model, find the average number of observations one needs to observe all the 625 distinct song elements. Is this number larger than 5500? Do you think 5500 is a sufficiently large sample to observe all the 625 distinct song elements?

2.5 Conditional expectation

To compute the expectation of a discrete random variable X, we multiply the values of X with the corresponding probabilities, and add them all:

$$E[X] = \sum_x xP\{X = x\}.$$

Now imagine a situation where we observe that another random variable Y takes the value y, and $P\{X = x\}$ the probabilities change accordingly to $P\{X = x \,|\, Y = y\}$. If we update the probabilities $P\{X = x\}$ with these conditional probabilities in the definition of $E[X]$, we obtain the conditional expectation of X given $Y = y$:

$$E[X \,|\, Y = y] = \sum_x xP\{X = x \,|\, Y = y\}.$$

Example 2.5 Roll two dice independently and let X_1 be the outcome of the first die, X_2 the second die, and X the sum of the two, that is, $X = X_1 + X_2$. Let's compute $E[X]$:

$$E[X] = E[X_1] + E[X_2] = 2(1(1/6) + 2(1/6) + 3(1/6) + 4(1/6) + 5(1/6) + 6(1/6)) = 7.$$

Let's see how this expectation changes if we were told that the outcome of the first die is 4, that is, we want to compute $E[X \,|\, X_1 = 4]$:

$$
\begin{aligned}
E[X \,|\, X_1 = 4] &= \sum_x xP\{X = x \,|\, X_1 = 4\} \\
&= 5P\{X = 5 \,|\, X_1 = 4\} + 6P\{X = 6 \,|\, X_1 = 4\} + 7P\{X = 7 \,|\, X_1 = 4\} + 8P\{X = 8 \,|\, X_1 = 4\} \\
&\quad + 9P\{X = 9 \,|\, X_1 = 4\} + 10P\{X = 10 \,|\, X_1 = 4\} \\
&= 5P\{X_2 = 1\} + 6P\{X_2 = 2\} + 7P\{X_2 = 3\} + 8P\{X_2 = 4\} + 9P\{X_2 = 5\} + 10P\{X_2 = 6\} \\
&= 15/2.
\end{aligned}
$$

The expectation of X can be recovered from the conditional expectations by a weighted average.

Lemma 2.7 *For discrete random variables X and Y, we have*

$$E[X] = \sum_y E[X \,|\, Y = y]P\{Y = y\}.$$

Proof The right-hand side simplifies as

$$
\sum_y P\{Y = y\} \sum_x xP\{X = x \,|\, Y = y\} = \sum_y P\{Y = y\} \sum_x x\frac{P\{X = x, Y = y\}}{P\{Y = y\}}
$$

$$
= \sum_y \sum_x xP\{X = x, Y = y\}
$$

$$= \sum_x x \sum_y P\{X = x, Y = y\}$$

$$= \sum_x x P\{X = x\}$$

$$= E[X].$$

Just like the expectation operation (see Note 2.1), the *conditional expectation* operation is linear.

Note 2.2 (Conditional expectation is linear) $E[\sum_{i=1}^n X_i \mid Y = y] = \sum_{i=1}^n E[X_i \mid Y = y]$.

In the previous example, we computed $E[X \mid X_1 = 4] = 15/2$. Similarly we could calculate quantities like $E[X \mid X_1 = 1], E[X \mid X_1 = 2], \ldots$. This motivates the following definition.

Definition 2.8 (Conditional expectation random variable) Let X, Y be discrete random variables. We define a new random variable denoted by $E[X|Y]$ as follows: if Y takes the value y, then $E[X|Y]$ takes the value $E[X|Y = y]$.

Notice that $E[X|Y]$ is defined as a function of Y: we know its value when we know the value Y attains.

Example 2.6 In Example 2.5 we computed $E[X \mid X_1 = 4] = 15/2$. Now we want to find $E[X \mid X_1]$. We have

$$E[X \mid X_1 = w] = \sum_x x P\{X = x \mid X_1 = w\} = \sum_{x=w+1}^{w+6} x P\{X = x \mid X_1 = w\} = \sum_{x=w+1}^{w+6} \frac{x}{6} = w + \frac{7}{2}.$$

Therefore, $E[X \mid X_1] = X_1 + \frac{7}{2}$.

Here's an interesting question. Since $E[X|Y]$ is itself a random variable, we can think about its expectation—so what is $E[E[X|Y]]$ equal to? Let's answer this for the previous example:

$$E[E[X|X_1]] = E[X_1 + 7/2] = E[X_1] + 7/2 = 7/2 + 7/2 = 7$$

which is interesting because $E[X]$ is also 7 (see Example 2.5 for that calculation). It turns out that this is always true.

Theorem 2.3 *For random variables X and Y, we have*

$$E[E[X|Y]] = E[X].$$

Proof Simply observe

$$E[E[X|Y]] = \sum_y E[X \mid Y = y] P\{Y = y\} = E[X],$$

the last equality following from Lemma 2.7. $\qquad\qquad\square$

Example 2.7 A rat is trapped in a maze and has two directions to choose from. If it goes to the right, then it will get out of the maze after 2 minutes of travel. If it goes to the left, it will wander around for 4 minutes before it comes back to the same location. Assume that the rat picks a direction at random each time. What is the expected travel time before it leaves the maze?

Let X be the time in the maze, and Y denote the direction. We will compute $E[X]$ by conditioning X on Y and using Lemma 2.7:

$$E[X] = E[X|Y = \text{right}]P\{Y = \text{right}\} + E[X|Y = \text{left}]P\{Y = \text{left}\}$$
$$= 2(1/2) + (4 + E[X])(1/2)$$
$$= 1 + 2 + E[X]/2.$$

Solving for $E[X]$, we get $E[X] = 6$.

2.5.1 Computing probabilities by conditioning

Suppose we want to compute the probability of some event E and there is a random variable Y such that if the outcome of Y is known, then $P(E)$ can be computed easily. In other words, we assume we can compute $P\{E \mid Y = y\}$ for any y. Can we obtain $P(E)$ from these conditional probabilities? The answer is yes, and it is nothing but a special case of Lemma 2.7:

$$P(E) = \sum_y P\{E \mid Y = y\}P\{Y = y\}.$$

Example 2.8 Let's revisit Example 2.4 but this time assume there are three types of army corps with different risks. Three Poisson random variables with parameters $\lambda = 0.7, 1.5, 3.0$ are used to model the number of deaths in each corp. Find the probability that there are two deaths in an army corp, if the army corp is picked at random with equal probabilities from the three.

Let T be the type of the army corp. $T = 1, 2, 3$ correspond to the corps with Poisson parameters $0.7, 1.5, 3.0$, respectively. Let X be the number of deaths. Then, by conditioning, we get the answer:

$$P\{X = 2\} = P\{X = 2 \mid T = 1\}P\{T = 1\} + P\{X = 2 \mid T = 2\}P\{T = 2\}$$
$$+ P\{X = 2 \mid T = 3\}P\{T = 3\}$$
$$= \left(e^{-0.7}\frac{(0.7)^2}{2!}\right)\frac{1}{3} + \left(e^{-1.5}\frac{(1.5)^2}{2!}\right)\frac{1}{3} + \left(e^{(-3.0)}\frac{(3.0)^2}{2!}\right)\frac{1}{3}$$
$$\approx 0.20.$$

Oftentimes, we want to know the probability that a random variable X is larger than a certain value, but we do not know enough about the random variable to compute this exactly. For example, let X be the number of snaps you get on Snapchat in a day (or, number of daily likes on social media). Clearly X varies from day to day, so it makes sense to think of it as a random variable. Suppose you look at the snaps you got every day for the last month, and found the daily average to be 30. Here is the burning question: what are the chances you will wake up one day to see you received at least 1000 snaps the previous day? We can get a rough estimate for this probability using Markov's inequality.

Theorem 2.4 (Markov's inequality) *If X is a random variable that takes on only nonnegative values, and k is some positive number, then*

$$P\{X \geq k\} \leq \frac{E[X]}{k}. \tag{2.5}$$

Going back to the Snapchat example, since the daily average snaps is 30, we put $E[X] = 30$, and set $k = 1000$ to find

$$P\{X \geq 1000\} \leq \frac{30}{1000} = 0.03.$$

Therefore the probability of getting at least 1000 snaps is at most 3%.

2.5.2 Martingales

In probability, we often study collections of random variables, as opposed to a single random variable at a time. We will see such examples in Chapters 4 and 5. Here we will discuss a particular sequence of random variables called a **martingale**.

Let $X_1, X_2, \ldots, X_n, \ldots$ be a sequence of discrete random variables. You can think of these random variables as random quantities you observe at time 1, 2, etc. We say this sequence is a martingale, if for any n, we have

$$E[X_{n+1} \mid X_n = x_n, X_{n-1} = x_{n-1}, \ldots, X_1 = x_1] = x_n. \tag{2.6}$$

In other words, if we observe that the random variables X_1 through X_n took the values x_1 through x_n, then the conditional expectation of X_{n+1}, a random variable we have not observed yet, equals x_n. At first glance, the definition of a martingale may not look very interesting. We should discuss some examples to get more insight.

A "fair game" is a game of chance where the expected value of one's earnings is zero. For example, imagine a game you play with a friend, where you flip a coin and your friend pays you \$1 if the coin lands heads. If the coin lands tails you pay your friend \$1. The expected value of your earnings is $\frac{1}{2} \times 1 + \frac{1}{2} \times (-1)$, which is zero. This is called a fair game because it is not biased against you or your friend.

Now imagine you play a fair game repeatedly. Let Y_n denote your earnings from the nth game. We will keep track of the total capital you have accumulated as you play the game: let X_n be the total capital at time n. Consider X_{n+1}, your total capital at time $n + 1$, given your past total capital $X_n = x_n, X_{n-1} = x_{n-1}, \ldots, X_1 = x_1$. We can write X_{n+1} as

$$X_{n+1} = x_n + Y_{n+1}.$$

In other words, total capital at time $n + 1$ equals total capital at time n (note that the capital at time $n − 1$ and earlier are irrelevant) plus the earnings from the $(n + 1)$th game. Therefore, the possible values for X_{n+1} are either $x_n + 1$ or $x_n − 1$ with probability $1/2$, since Y_{n+1}, the earnings from the $(n + 1)$th game, is either 1 or −1 with probability $1/2$. Then the expected value of X_{n+1} is $(x_n + 1)/2 + (x_n − 1)/2 = x_n$. In summary, we have proved

$$E[X_{n+1} \mid X_n = x_n, X_{n-1} = x_{n-1}, \ldots, X_1 = x_1] = x_n + E[Y_{n+1}] = x_n,$$

and thus the total capital, when one plays a fair game, is a martingale.

It will be instructive to simulate a fair game, and plot the capital as a function of time. We start with an initial capital of \$10, so that $X_1 = 10 + Y_1$. Note that $E[X_1] = 10 + E[Y_1] = 10$, since the game is fair and $E[Y_1] = 0$. If our capital becomes negative, we continue playing the game by borrowing money, so a negative capital means we owe that much. We play the game 500 times, and plot the capital $X_1, \ldots, X_{500}$. We then repeat this ten times. Figure 2.6 plots the resulting ten curves. The initial capital of \$10 is plotted as a horizontal line.

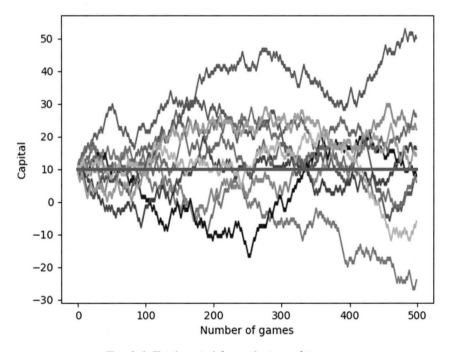

Fig. 2.6: Total capital from playing a fair game

Figure 2.6 shows that we can get lucky, and have a capital larger than \$10 most of the time, but of course we can also get unlucky and be in debt. However, there is an interesting symmetry in the figure. If you pick a fixed time on the x-axis, and eyeball the ten values for the capital at that time, you will see that the values look like evenly spread around the horizontal line. In other words, it looks like $E[X_n] = E[X_1] = 10$ for any n. Although some curves rise and some fall, on the average, the capital is flat.

Our intuition is correct, and it has a simple proof. Using Lemma 2.7 we write

$$E[X_2] = \sum_{x_1} E[X_2 \mid X_1 = x_1] P\{X_1 = x_1\} = \sum_{x_1} x_1 P\{X_1 = x_1\} = E[X_1],$$

where we used the definition of the martingale to replace $E[X_2 \mid X_1 = x_1]$ by x_1. Similarly we can prove $E[X_3] = E[X_2]$, and continue in this manner until we prove $E[X_n] = E[X_{n-1}]$.[4]

2.5.2.1 Wright–Fisher model

Every gene has two variants, called alleles. Consider a gene with two alleles, **A** and **a**, in a population. Each individual in the population has a genotype of either **AA**, **aa**, or **Aa**. Now consider the genotype of the entire population, and let p be the proportion of allele **A** in the gene pool. As individuals die, and offsprings arise, the proportion p will change, a phenomenon called *genetic drift* in population genetics.

The Wright–Fisher model is a model for the evolution of the frequency of alleles across generations. It ignores mutation, natural selection, and other external factors, and only considers the random sampling of the alleles as a result of reproduction. The model assumes generations do not overlap. Assume we have a population of size N. Then all the N individuals in the first generation will be replaced by N offsprings in the second generation, and so on. Note that since each individual has two alleles, the total number of alleles in the population is $2N$.

Let's describe how reproduction works in the Wright–Fisher model. The genotype of an offspring is determined as follows. Assume the number of allele **A** in generation n is k. Pick an allele at random, uniformly from the space of $2N$ alleles, k of which are allele **A**, and the rest are allele **a**. Let's say we picked allele **a**. Now pick another allele at random, from the same space. Let's say this time we got allele **A**. Then the offspring's genotype is **Aa**.

Let X_n denote the number of **A** alleles at generation n. We want a model for X_{n+1}, based on the information we have from generation n. The Wright–Fisher model assumes X_{n+1} is a random variable satisfying the following conditional probabilities

$$P\{X_{n+1} = i \mid X_n = k\} = \binom{2N}{i} \left(\frac{k}{2N}\right)^i \left(1 - \frac{k}{2N}\right)^{2N-i}, \tag{2.7}$$

where i, k are integers between 0 and $2N$.

---

[4] Readers familiar with mathematical induction should try making this last statement rigorous using induction.

Observe that this conditional probability is the probability mass function for a binomial random variable with parameters $2N$ (number of trials) and $\frac{k}{2N}$ (probability of success). Think about picking allele **A** as "success", and picking allele **a** as "failure". Since $X_n = k$, there are k alleles **A** in generation n, and thus the probability of success is $\frac{k}{2N}$. The number of successes is the number of alleles **A** in generation $n + 1$, which is X_{n+1}, and hence Eq. (2.7) follows.

Here is the main result we have been working toward: $X_1, X_2, \ldots$ is a martingale. The proof is simple; just recall that the expected value of a binomial random variable is the product of its parameters, and observe

$$E[X_{n+1} \mid X_n = x_n, \ldots, X_1 = x_1] = 2N \left(\frac{x_n}{2N} \right) = x_n.$$

What is the significance of this result? The frequency of an allele may rise and fall in a population across generations due to chance, but the expected value of the frequency remains the same as the first generation, since we have proved earlier that $E[X_n] = E[X_1]$ for a martingale. The random sampling of genes in the Wright–Fisher model does not favor one allele over another. In other words, in this model the genetic drift has no direction, and only governed by chance. Recall that the Wright–Fisher model ignores external factors such as natural selection, which could very well favor one allele over another, and then the martingale property would no longer be true.

2.6 Project 7: A professor's trick

A probability professor who likes to play tricks on her students tells her students in the first lecture that she has to step out of the classroom for 10 minutes. She tells them to flip a coin 100 times and record the outcomes on a piece of paper while she is away. Now flipping a coin 100 times is no doubt boring, and a few students decide to be a little creative, and jot down some heads and tails on paper confident that they can perfectly simulate the outcomes of a real coin in their minds! How wrong they are! Upon her return, the professor asks how many students had 4 consecutive heads or tails. There are 10 students in the class, and 6 students raise their hands. The professor looks at the students who did not raise their hands, and says: "I am 95% confident at least one of you did not really flip your coins!".

Your mission, if you choose to accept it, is to investigate how the professor identified the jokers in the classroom, and save her future students from further embarrassment!

Let's generalize the problem a little before solving it. Consider a random experiment with n equally likely outcomes. We repeat this random experiment, independently, until the same outcome occurs k consecutive times. Let N_k denote the number of trials that is needed to observe the same outcome k consecutive times. We want to find $E[N_k]$.

1. First, explain why the following equation is correct (comment on how the equation is derived, and why the limits for the index j are $k-1$ and ∞):

$$E[N_k] = \sum_{j=k-1}^{\infty} E[N_k \mid N_{k-1} = j]P\{N_{k-1} = j\}.$$

   Then show that the right-hand side of the equation can be simplified as

$$E[N_k] = \frac{1}{n}(E[N_{k-1}] + 1) + \frac{n-1}{n}E[N_{k-1}] + \frac{n-1}{n}E[N_k].$$

   Finally, solve the above equation algebraically to derive the recursion

$$E[N_k] = nE[N_{k-1}] + 1 \tag{2.8}$$

   for $k = 2, 3, \ldots$.
2. Prove that the formula $E[N_k] = \frac{n^k - 1}{n - 1}$ satisfies the recursion (2.8), where $k = 1, 2, \ldots$.
3. Use Markov's inequality (Theorem 2.4) to find an upper bound for the probability $P\{N_k \geq a\}$.
4. Back to the professor's trick: put $n = 2, k = 4$, and shed light into the trick using your answers to parts 1 and 2.

Chapter 3
Continuous Random Variables

In Chapter 2, we discussed discrete random variables. These random variables took on finite or countably many values. Here we will study random variables whose set of possible values is uncountable. For example, the price of a stock with its erratic ups and downs can be modeled as a random variable, and since in principle the price can be any positive number, modeling it using a continuous random variable is a sensible strategy. Another example is the waiting time for the first customer to enter a store, if you like, your neighborhood Starbucks. The waiting time can be any positive number (if we could measure it with sufficient precision), so a continuous random variable could be a good model.

Definition 3.1 (Continuous random variable) We say a random variable X is continuous if there exists a nonnegative function f with the property that for any set B of real numbers

$$P\{X \in B\} = \int_B f(x)dx.$$

The function f is called the **probability density function** (pdf) of X. The support of X, denoted by $\mathcal{S}$, is the set of all real numbers x such that $f(x) > 0$.

In the case of a discrete random variable, to find $P\{X \in B\}$ we would simply add all the *probability mass function* values $\sum_{b \in B} f(b) = \sum_{b \in B} P\{X = b\}$. We can compute this summation if B is finite, and sometimes when it is countably infinite. However, if B is uncountable, this approach no longer works because we cannot add uncountably many numbers. Therefore, in the case of continuous random variables, we need another approach. This approach is explained by Definition 3.1: we seek functions that give us probabilities when we *integrate* them, instead of *summing* them. These functions are called probability density functions.

Let's go back to the example of the waiting time for the first customer to enter a store, and denote this waiting time by X. It turns out that often a *good model* for the waiting time for various things is a continuous random variable with the following pdf

$$f(x) = \lambda e^{-\lambda x}, \ 0 \le x < \infty,$$

© The Editor(s) (if applicable) and The Author(s), under exclusive
license to Springer Nature Switzerland AG 2020
G. Ökten, *Probability and Simulation*, Springer Undergraduate Texts
in Mathematics and Technology, https://doi.org/10.1007/978-3-030-56070-6_3

where λ is a positive real number that varies based on the problem. Let's elaborate on what is meant by a continuous random variable being a good model for some observations.

Table 3.1 reports the time (in hours) spent to see a physician in emergency rooms in the U.S. in 2016[1]. For example, the entry 14,567 means as many patients waited between 1 and 2 hours to be seen by a doctor. The total number of visits in the table is 125,915.

Waiting time	$t < 1$	$1 \le t < 2$	$2 \le t < 3$	$3 \le t < 4$	$4 \le t < 6$	$t \ge 6$
Number of visits	102,796	14,567	4715	1742	1345	750

Table 3.1: Waiting time (in hours) at emergency room visits in the U.S. in 2016. The mean waiting time is 42 minutes.

Figure 3.1 superimposes two plots. One is a bar graph where the heights of the bars are the relative frequencies of the number of visits in Table 3.1 and the widths, corresponding to the time intervals, are about one (except for the interval $[4, 6]$). The leftmost bar indicates about 80% (102,796/125,915) of patients waited for less than an hour. The x-axis shows the time in hours. The continuous curve in Figure 3.1 is the plot of $f(x) = 1.4e^{-1.4x}$: this is the pdf of a continuous random variable, call it X, discussed earlier with $\lambda = 1.4$. Note how well the continuous curve matches the shape of the bar graph: this observation is precisely what is meant when we say a continuous random variable is a good model for some data.

If we accept that X, the continuous random variable with pdf $f(x) = 1.4e^{-1.4x}$, $0 \le x < \infty$, is a good model for the waiting time data, then we can answer questions about the data using it. For example, we already know from the data that the relative frequency of visits that require less than an hour waiting is about 80%. What does the model say this relative frequency, or more precisely, the probability $P\{0 < X < 1\}$, is? From Definition 3.1, the probability is obtained by integrating the pdf from 0 to 1:

$$P\{0 < X < 1\} = \int_0^1 1.4e^{-1.4x}\,dx = 0.75,$$

which is fairly close to 80%. This illustrates the heights of the bars (which is the same as the area when the width of the bar is one) in the bar plot, correspond to the areas under the smooth pdf.

Now let's get some new information about the data. You may have noticed that in Table 3.1, the time intervals do not have uniform spacing. For example, although the first four intervals have width one, the fifth one, $4 \le t < 6$, has width two. What if we needed to know the relative frequency of visits that had a waiting time between 4 and 5 hours? The data does not have this information, but we can compute the probability $P\{4 < X < 5\}$ using our model for an approximate answer:

---

[1] National Hospital Ambulatory Medical Care Survey: 2016 Emergency Department Summary Tables, https://www.cdc.gov/nchs/data/nhamcs/web_tables/2016_ed_web_tables.pdf

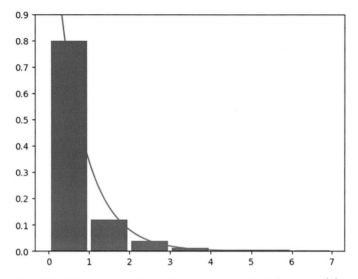

Fig. 3.1: Waiting time relative frequencies and $f(x) = 1.4e^{-1.4x}$

$$P\{4 < X < 5\} = \int_4^5 1.4e^{-1.4x}\,dx = 0.0028,$$

or about 0.3%.

Let's make a few more observations about the random variable X with pdf $f(x) = 1.4e^{-1.4x}, 0 \le x < \infty$. From the fact that the pdf is defined on $[0, \infty)$, we deduce the support of X is the interval $[0, \infty)$. We can extend the definition of $f(x)$ to the set of all real numbers $\mathbb{R}$ by setting f to zero outside the support:

$$f(x) = \begin{cases} 1.4e^{-1.4x}, & 0 \le x < \infty \\ 0, & \text{otherwise.} \end{cases} \tag{3.1}$$

Here is one advantage of extending the pdf of any random variable X to $\mathbb{R}$: we can now write

$$\int_{-\infty}^{\infty} f(x)\,dx = P\{X \in \mathcal{S}\} = 1,$$

where $\mathcal{S}$ is the support of the random variable.

There is another important function associated with a random variable: its **cumulative distribution function**.

Definition 3.2 (Cumulative distribution function) The cumulative distribution function (cdf) of a continuous random variable X is defined by

$$F(x) = P\{X \le x\} = \int_{-\infty}^{x} f(t)\,dt.$$

We now compute the cdf of the random variable X we discussed earlier. From (3.1), we observe first that $F(x) = 0$ if $x < 0$. If $x > 0$, then

$$F(x) = \int_0^x 1.4e^{-1.4t} dt = -e^{-1.4t}\Big|_0^x = 1 - e^{-1.4x}.$$

Hence

$$F(x) = \begin{cases} 0, & -\infty < x < 0 \\ 1 - e^{-1.4x}, & x > 0 \end{cases}. \tag{3.2}$$

For example, $F(3) = 1 - e^{-1.4 \times 3} = 0.985$, which means $P\{X \leq 3\} = 0.985$.

Remark 3.1 The relationship between the pdf and cdf of a continuous random variable follows from the fundamental theorem of calculus:

$$F(x) = \int_{-\infty}^x f(t)dt \Rightarrow \frac{d}{dx}F(x) = f(x). \tag{3.3}$$

Of course, the derivative of F will be equal to f, provided F' exists in the first place.

The expectation and variance of a continuous random variable is defined similar to those of a discrete random variable where summations are replaced by integrals, and the probability mass functions are replaced by probability density functions.

Definition 3.3 Let X be a continuous random variable with pdf $f(x)$. Then

• The expected value of X is

$$\mu = E[X] = \int_{-\infty}^{\infty} xf(x)dx,$$

and the expected value of a function of X is

$$E[g(X)] = \int_{-\infty}^{\infty} g(x)f(x)dx.$$

• The variance of X is

$$\sigma^2 = Var(X) = E[(X - \mu)^2] = \int_{-\infty}^{\infty} (x - \mu)^2 f(x)dx,$$

with the familiar identity

$$\sigma^2 = E[X^2] - E[X]^2$$

being still valid.

The properties of the expected value and variance for discrete random variables are still valid in the case of continuous random variables. Here are some we will frequently use:

• $E[X + Y] = E[X] + E[Y]$
• $E[aX] = aE[X]$

- $Var(X + Y) = Var(X) + Var(Y)$, if X and Y are independent
- $Var(aX) = a^2 Var(X)$,

where a is a constant. The first and third identities generalize to any number of random variables in a straightforward way.

Example 3.1 The density function of X is given by

$$f(x) = \begin{cases} 1/2, & \text{if } 0 < x < 2 \\ 0, & \text{otherwise} \end{cases}.$$

Find (i) the cdf of X, (ii) $P\{-1 < X < 1.5\}$, (iii) μ, σ^2.

Solution 3.1 • The cdf is

$$F(x) = \int_{-\infty}^{x} f(t)dt = \begin{cases} \int_{-\infty}^{x} 0 dt = 0, \text{ if } x < 0 \\ \int_{0}^{x} (1/2)dt = x/2, \text{ if } 0 \le x < 2 \\ \int_{0}^{2} (1/2)dt + \int_{2}^{x} 0 dt = 1, \text{ if } x \ge 2 \end{cases}$$

- $P\{-1 < X < 1.5\} = \int_{-1}^{0} 0 dt + \int_{0}^{1.5} (\frac{1}{2})dt = 0 + \frac{1}{2} \times (1.5) = \frac{3}{4}.$
- The mean is

$$\mu = \int_{0}^{2} x(1/2)dx = \frac{1}{2}\frac{x^2}{2}\Big|_{0}^{2} = 1.$$

The variance is

$$\sigma^2 = \int_{0}^{2} (x-1)^2 (1/2)dx = \frac{1}{2}\frac{(x-1)^3}{3}\Big|_{0}^{2} = \frac{1}{6} - \left(\frac{-1}{6}\right) = \frac{2}{6}.$$

Remark 3.2 Recall that for a discrete random variable X, the probability that X takes the value k is given by $f(k)$ where f is the probability mass function, that is,

$$P\{X = k\} = f(k).$$

However, for a continuous random variable X with probability density function f, we have

$$P\{X = k\} = \int_{k}^{k} f(x)dx = 0,$$

therefore $P\{X = k\}$ cannot be the same as $f(k)$! For continuous random variables, the relationship between $P\{X = k\}$ and $f(k)$ is more subtle:

$$P\left\{k - \frac{\varepsilon}{2} \le X \le k + \frac{\varepsilon}{2}\right\} = \int_{k-\varepsilon/2}^{k+\varepsilon/2} f(t)dt \approx \varepsilon f(k),$$

where the integral is approximated by the midpoint rule. Intuitively, this formula shows $f(k)$ is approximately the probability that X takes values within a small interval around k divided by the length of the interval.

3.1 Uniform random variables

We say X is a **uniform random variable** on $(0, 1)$ (or, X has a **uniform distribution** on $(0, 1)$) if its pdf is given by

$$f(x) = \begin{cases} 1, & \text{if } 0 < x < 1 \\ 0, & \text{otherwise.} \end{cases}$$

Observe:

1. Since $f(x) > 0$ only when $x \in (0, 1)$, X takes values only between 0 and 1.
2. Since $f(x)$ is constant when $x \in (0, 1)$, X is just as likely to be near any value in $(0, 1)$ as any other value.
3. The probability that X is in a particular subinterval (α, β) is the length of the interval, since

$$P\{\alpha < X < \beta\} = \int_\alpha^\beta 1 dx = \beta - \alpha.$$

Now we give the definition of uniform distribution on any interval (a, b).

Definition 3.4 We say X is a uniform random variable on the interval (a, b), and denote it by $U(a, b)$, if its pdf is given by

$$f(x) = \begin{cases} \frac{1}{b-a}, & \text{if } a < x < b \\ 0, & \text{otherwise.} \end{cases}$$

The cdf of X is

$$F(x) = \begin{cases} \int_{-\infty}^x 0 dt, & \text{if } x < a \\ \int_{-\infty}^a 0 dt + \int_a^x \frac{1}{b-a} dt, & \text{if } a \leq x < b \\ \int_{-\infty}^a 0 dt + \int_a^b \frac{1}{b-a} dt + \int_b^x 0 dt, & \text{if } x \geq b \end{cases} = \begin{cases} 0, & \text{if } x < a \\ \frac{x-a}{b-a}, & \text{if } a \leq x < b \\ 1, & \text{if } x \geq b \end{cases}.$$

Theorem 3.1 *Let X be $U(a, b)$. Then*

$$E[X] = \frac{a+b}{2}, \quad Var(X) = \frac{(b-a)^2}{12}.$$

Proof Let's prove the first statement.

$$E[X] = \int_a^b \frac{x}{b-a} dx = \frac{1}{b-a} \frac{x^2}{2} \Big|_b^a = \frac{1}{b-a} \frac{b^2 - a^2}{2} = \frac{b+a}{2}.$$

3.1.1 Strong law of large numbers

Let $X_1, X_2, \ldots, X_N$ be independent random variables with the same distribution, and consider their sample mean

$$S_N = \frac{X_1 + X_2 + \ldots + X_N}{N}.$$

Note that S_N itself is a random variable. We want to generate some values from S_N to get some insight on the random variable. We will run some Julia simulations to do that. Assume the common distribution of the X_n is $U(0, 1)$, the uniform distribution on $(0, 1)$. To generate values from S_N, we will generate N random numbers $x_1, \ldots, x_N$ from $U(0, 1)$, and compute $s_N = \frac{x_1 + x_2 + \ldots + x_N}{N}$. This gives us one value from the random variable S_N. We will then repeat this five times and obtain five values from S_N: $s_N^{(1)}, \ldots, s_N^{(5)}$. Note that we are using capital letters for random variables, and small letters for specific real numbers generated.

The Julia function **rand(N)** generates N random numbers from $U(0, 1)$, and the function **sum** computes the sum of the elements of the array the function is applied to. So, for example, **sum(rand(N))** returns the sum of the N random numbers. Here is the full Julia code.

```
In [1]: function sMean(N)
            means=zeros(5)
            for i in 1:5
                means[i]=sum(rand(N))/N
            end
            means
        end
```

```
Out[1]: sMean (generic function with 1 method)
```

Let's pick $N = 10$ and compute five values from $s_N = \frac{x_1 + x_2 + \ldots + x_N}{N}$:

```
In [2]: sMean(10)
```

```
Out[2]: 5-element Array{Float64,1}:
        0.5372020540485136
        0.5274863532773761
        0.6064548065514324
        0.5512406449209623
        0.3702736760827891
```

We see numbers ranging from 0.3 to 0.6. Let's increase N to 10000 and see what happens:

```
In [3]: sMean(10000)
```

```
Out[3]: 5-element Array{Float64,1}:
        0.5018929310023443
        0.49494789673660033
        0.49821360176509344
        0.5016461189671038
        0.5064912466112997
```

Interestingly, the numbers seem to be all around 0.5. In other words, S_N seems to be converging to 0.5, when we increase N. Is there anything special about 0.5? Well, it is the common expected value of the random variables $X_1, \ldots, X_N$! Recall that each X_n is $U(0, 1)$, and the expected value of $U(0, 1)$ is 1/2.

If we used a distribution other than $U(0, 1)$, and computed S_N in a similar way, would we still see the values from S_N converging to the expected value of the distribution? The answer is yes, and this result is a famous theorem in probability theory known as the **strong law of large numbers**. More precisely, the theorem states

$$S_N = \frac{X_1 + X_2 + \ldots + X_N}{N} \to E[X]$$

with *probability one*, as $N \to \infty$, where X_n's are independent, and they have the same distribution as X. The phrase *probability one* can be best explained in more advanced textbooks on probability. Intuitively it means the following: if we repeated the above Julia calculations infinitely many times, the proportion of times when S_N *does not converge* to $E[X]$ will tend to zero. A proof of the theorem can be found in advanced textbooks, such as Billingsley [5].

3.2 Project 8: Monte Carlo integration

Estimation of the integral $I = \int_{(0,1)^s} g(x)dx$ is a classical problem in numerical analysis known as the quadrature problem. The Monte Carlo integration refers to algorithms that estimate this integral using random numbers. Here we will discuss two such methods.

Hit-or-miss Monte Carlo

Let g be a function from $(0, 1)$ to $(0, 1)$. We want to estimate $I = \int_0^1 g(x)dx$, which is the area under the graph of g; see Figure 3.2.

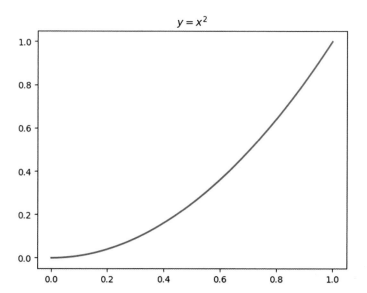

Fig. 3.2: Finding the area under $y = x^2$ with hit-or-miss Monte Carlo

Now suppose we generate N uniform random numbers from the unit square, and count how many fall under the graph of g—say this number is S. Then the area under g is approximately $\frac{S}{N}$, that is

$$I = \int_0^1 g(x)dx \approx \frac{S}{N}.$$

1. Plot $y = x^2$ on a board, enclosed within a unit square, and take turns to throw 40 darts at the unit square.[2] Count S, N and compute $\frac{S}{N}$. How close is it to the exact area $\int_0^1 x^2 dx = \frac{1}{3}$?
2. Let's analyze this method theoretically. Let

$$X_i = \begin{cases} 1 \text{ if the } i\text{th dart falls under the graph} \\ 0 \text{ otherwise.} \end{cases}$$

   a. What is the probability $P\{X_i = 1\}$ for any i?
   b. Write S in terms of X_i. What is the distribution of the random variable S?
   c. Find the expectation $E[S/N]$.
   d. Find the variance $Var(S/N)$.

Sample mean Monte Carlo

Let $U_1, U_2, \ldots, U_N, \ldots$ be independent uniform random variables on $(0, 1)$. Then the sample mean

$$\frac{1}{N} \sum_{i=1}^{N} g(U_i) = \frac{g(U_1) + \ldots + g(U_N)}{N}$$

is an approximation to the expectation

$$E[g(U)] = \int_0^1 g(x)dx = I,$$

with the error of the approximation converging to 0 (with probability 1) as $N \to \infty$. This follows from the strong law of large numbers we discussed earlier.

1. Write a Julia code to approximate $\int_0^1 e^{x^2} dx$ using sample mean Monte Carlo. The code should take N as the input, and return $(g(U_1) + \ldots + g(U_N))/N$ as the output, where $g(x) = e^{x^2}$. The U_i's will be replaced by random numbers in your code. Recall that the function **rand()** generates a random number from the uniform distribution on $(0, 1)$. Run your code with $N = 1000$ and $N = 10,000$. Compare your results with the approximation WolframAlpha gives for the integral.
2. The sample mean Monte Carlo generalizes to higher dimensions in a straightforward way. If g is a function on $(0, 1)^s$, then

$$\frac{1}{N} \sum_{i=1}^{N} g(U_i) \approx E[g(U)] = \int_{(0,1)^s} g(x)dx = I$$

---

[2] YSP students at Florida State University shoot a Nerf gun aiming at the unit square drawn on the blackboard.

where U_i is an s-dimensional vector, with components independent uniform random variables on $(0, 1)$. Write a Julia code to estimate $\int_0^1 \int_0^1 e^{(x+y)^2} dx dy$.

3. Sample mean Monte Carlo can be generalized to estimate integrals defined on a finite interval (a, b). Prove that

$$\int_a^b g(x)dx \approx \frac{b-a}{N} \sum_{i=1}^N g(a + (b-a)u_i)$$

where u_i are uniform random numbers from $(0, 1)$. (Hint: Use the substitution $x = \frac{y-a}{b-a}$ to turn the integral to one with domain $(0, 1)$.)

4. Sample mean Monte Carlo is a better method than hit-or-miss Monte Carlo. These methods estimate $\int_0^1 g(x)dx$ by computing

- S/N in hit-or-miss MC
- $\frac{1}{N} \sum_{i=1}^N g(U_i)$ in sample mean MC.

Claiming that sample mean Monte Carlo is better than hit-or-miss Monte Carlo means the variance of its estimates is smaller than the variance of the estimates of hit-or-miss, that is

$$Var\left(\frac{1}{N} \sum_{i=1}^N g(U_i)\right) \leq Var(S/N).$$

Prove this inequality.

3.3 Exponential and normal random variables

When we first introduced the definition of a continuous random variable in Chapter 3, we discussed an example of a random variable whose pdf was an exponential function. Such random variables are ubiquitous in probability theory, and they are called **exponential random variables.**

Definition 3.5 We say X is an exponential random variable with parameter $\lambda > 0$ and write $X \sim \exp(\lambda)$ if its pdf is given by

$$f(x) = \begin{cases} \lambda e^{-\lambda x}, & \text{if } x \geq 0 \\ 0, & \text{otherwise} \end{cases}.$$

The cdf of $X \sim \exp(\lambda)$ is

$$F(x) = \begin{cases} 1 - e^{-\lambda x}, & \text{if } x \geq 0 \\ 0, & \text{otherwise} \end{cases};$$

we have seen a derivation of this in the aforementioned example (see Eq. (3.2)).

Some authors, including Julia, define the exponential random variable slightly differently. They write the pdf as $f(x) = \frac{1}{\theta} e^{-x/\theta}$. What we call λ corresponds to $1/\theta$ in this definition.

Theorem 3.2 *Let X be an exponential random variable with parameter $\lambda > 0$. Then $E[X] = \frac{1}{\lambda}$ and $Var(X) = \frac{1}{\lambda^2}$.*

Proof We will prove the first statement. From the definition of expectation, we have

$$E[X] = \int_0^\infty \lambda x e^{-\lambda x} \, dx,$$

where the integral is an improper integral. Using integration by parts, it is not too difficult to obtain the indefinite integral

$$\int x e^{-\lambda x} \, dx = -\frac{e^{-\lambda x}}{\lambda^2} - \frac{x e^{-\lambda x}}{\lambda}.$$

Then we can compute

$$E[X] = \int_0^\infty \lambda x e^{-\lambda x} \, dx = \lim_{a \to \infty} \lambda \int_0^a x e^{-\lambda x} \, dx$$

$$= \lambda \lim_{a \to \infty} \left[\left(-\frac{e^{-\lambda x}}{\lambda^2} \right)\Big|_0^a - \left(\frac{x e^{-\lambda x}}{\lambda} \right)\Big|_0^a \right] = \frac{1}{\lambda}.$$

Lemma 3.1 *If X is an exponential random variable with parameter λ, then $P\{X > x\} = e^{-\lambda x}$.*

Proof First note that the events $\{X > x\}$ and $\{X \leq x\}$ are complementary, and thus $P\{X > x\} + P\{X \leq x\} = 1$. Since $P\{X \leq x\} = F(x) = 1 - e^{-\lambda x}$, the conclusion follows. □

Example 3.2 Exponential random variables are often used to model the waiting time between successive changes, or waiting time until the first change. For example, the waiting time between consecutive customers arriving at a store can be modeled via an exponential random variable. Let W be the time in minutes between two student arrivals at FSU Starbucks between 12PM and 4PM. From watching students and collecting data, we think the average waiting time is 2.5 minutes. Find (i) $P\{W > 2\}$, (ii) $P\{W > 16 \,|\, W > 14\}$.

Solution 3.2 Since $E[W] = \frac{1}{\lambda} = 2.5$, we get $\lambda = 0.4$. Then

$$P\{W > 2\} = e^{-0.4(2)} = e^{-0.8} \approx 0.45,$$

and

$$P\{W > 16 \,|\, W > 14\} = \frac{P\{W > 16, W > 14\}}{P\{W > 14\}} = \frac{P\{W > 16\}}{P\{W > 14\}} = \frac{e^{-16(0.4)}}{e^{-14(0.4)}} = e^{-0.4(2)}$$
$$\approx 0.45,$$

where we used Lemma 3.1 to compute $P\{W > 14\}, P\{W > 16\}$.

Theorem 3.3 *The exponential random variable is memoryless, that is,*

$$P\{X > s + t \,|\, X > t\} = P\{X > s\}, \text{ for all } s, t \geq 0.$$

Proof Using the definition of conditional probability, we write

$$P\{X > s + t \,|\, X > t\} = \frac{P\{X > s + t, X > t\}}{P\{X > t\}}.$$

The event in the numerator, $\{X > s + t, X > t\}$, is the intersection of the events $\{X > s + t\}$ and $\{X > t\}$. Note that a number is greater than $s + t$ *and* t, if and only if it is greater that $s + t$. Think about these sets as rays on the real line, one extending from $s + t$ to infinity, and the other from t to infinity (note that both s, t are positive numbers) to realize that the intersection of the two sets is $\{X > s + t\}$. Therefore,

$$P\{X > s + t \,|\, X > t\} = \frac{P\{X > s + t, X > t\}}{P\{X > t\}} = \frac{P\{X > s + t\}}{P\{X > t\}} = \frac{e^{-\lambda(s+t)}}{e^{-\lambda t}}$$

where we used Lemma 3.1 in the last step. The expression on the right-hand side simplifies as $e^{-\lambda s}$ which equals $P\{X > s\}$ from Lemma 3.1, proving the theorem.

Where does the name "memoryless" come from in the statement of the previous theorem? Let X be the waiting time for the first customer to enter a store and you are the store manager. The conditional probability $P\{X > s + t \,|\, X > t\}$ is the probability that you will wait at least *s more* units of time for the first customer, knowing that nobody walked in during the first *t* units of time. The memoryless property says this probability is the same as $P\{X > s\}$, the probability that you would have waited for at least *s* units of time from the time you first opened the store. In other words, just because nobody walked in, say, in the first 10 minutes since you opened the store, there won't be a customer rushing into your store to make up for the missed time!

3.3.0.1 Exponential distribution in Julia

We start with loading the package **Distributions** which contains a large collection of probability distributions.

In [1]: `using Distributions`

In [2]: `using PyPlot`

Let's ask Julia for information on the exponential distribution.

In [3]: `?Exponential`

search: **Exponential Exponential**BackOff Double**Exponential exponent**

Out[3]:

`Exponential(`θ`)`

The *Exponential distribution* with scale parameter θ has probability density function

$$f(x;\theta) = \frac{1}{\theta}e^{-\frac{x}{\theta}}, \quad x > 0$$

```
Exponential()        # Exponential distribution with unit scale,
                     i.e. Exponential(1)
Exponential(b)       # Exponential distribution with scale b

params(d)            # Get the parameters, i.e. (b,)
scale(d)             # Get the scale parameter, i.e. b
rate(d)              # Get the rate parameter, i.e. 1 / b
```

External links

• Exponential distribution on Wikipedia

Recall that Julia does not follow Definition 3.5, but its modification where λ is replaced by $1/\theta$. Let's define *X* as an exponential random variable with $\theta = 2$:

In [4]: `X=Exponential(2)`

Out[4]: `Exponential{Float64}(`θ`=2.0)`

The pdf and cdf of X can be evaluated by simply using the Julia functions with the same name:

```
In [5]: pdf(X,0.1)
```

```
Out[5]: 0.475614712250357
```

```
In [6]: cdf(X,3)
```

```
Out[6]: 0.7768698398515702
```

Next we plot the pdf and cdf of X:

```
In [7]: xval=range(0,stop=10,length=100)
        y=pdf.(X,xval)
        z=cdf.(X,xval)
        plot(xval,y,label="PDF")
        plot(xval,z,label="CDF")
        legend(loc="upper right");
```

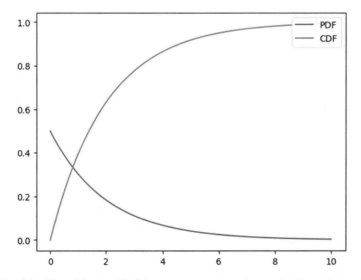

Fig. 3.3: The pdf and cdf of the exponential random variable with $\theta = 2$

3.3.0.2 Generating random numbers from exponential distribution

The function **rand(X)** generates a random number from the distribution X.

```
In [8]: rand(X)
```

```
Out[8]: 0.23795599865682912
```

Multiple random numbers from X can be obtained simply by

```
In [9]: rand(X,5)

Out[9]: 5-element Array{Float64,1}:
        1.352507567243558
        0.2585805359849165
        1.3950209909801357
        4.268198745211011
        0.7315039816318627
```

Below we generate 10000 random numbers and plot their relative frequency histogram in Figure 3.4. Notice how the histogram matches the shape of the exponential probability density function in Figure 3.3.

```
In [10]: hist(rand(X,10000),density=true);
```

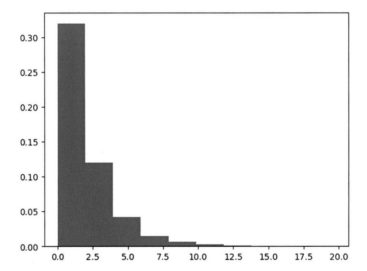

Fig. 3.4: 10000 random numbers from the exponential random variable with $\theta = 2$

Example 3.3 Many new businesses go bankrupt in a short amount of time, and increasingly smaller numbers survive as years pass by. If we examined the time-to-bankruptcy data for firms, would we see any interesting patterns? Fujiwara [9] (also see Aoyama et al. [1]) report data that consists of the number of days to bankruptcy of all the Japanese firms that went bankrupt in the year 1997.

There are 16,321 numbers in the data set; that's how many companies went bankrupt in 1997. The smallest number is 18, which means one company went bankrupt in just 18 days. The largest number is 47,206, which is about 130 years! So this company was founded in 1867, and went bankrupt in 1997. Figure 3.5 plots the relative frequency histogram for these 16,321 numbers.

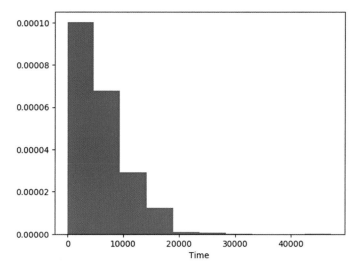

Fig. 3.5: Relative frequency histogram of time-to-bankruptcy

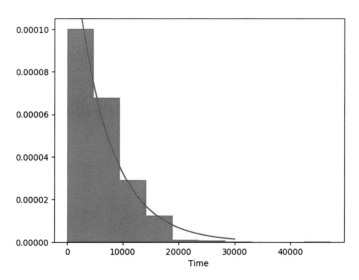

Fig. 3.6: Relative frequency histogram with pdf of $X \sim \exp(\theta = 6312)$

The shape of the histogram looks very much like the shape of the exponential probability density function we have seen before. Let's assume we are correct about this hunch, and that the data follows an exponential distribution with some parameter λ. But, what would λ be? Recall that if $X \sim \exp(\lambda)$, then $E[X] = 1/\lambda$. We can compute the mean of the data, which turns out to be about 6,312 days, and set $1/\lambda = 6312$, or $\lambda = 1/6312$. Figure 3.6 superimposes the probability density function of the random variable $X \sim \exp(1/6312)$ to the histogram. Visually, it is a great match!

If we agree that time-to-bankruptcy can be modeled using an exponential random variable with $\lambda = 1/6312$, we can extract lots of useful information from this model. For example, what are the chances that a firm that went bankrupt did so after 150 years (54750 days) of opening its doors? We cannot estimate this probability from the data, because the largest time in the data set is 130. However, using our model, we can compute

$$P\{X > 54750\} = 1 - P\{X \leq 54750\} = 1 - F(54750) = 0.00017,$$

which is about 0.02%, where F is the cdf for $X \sim \exp(1/6312)$. When we use this model, we should not forget that it is based on data from Japanese companies going bankrupt in a specific year, and be careful when we make extrapolations to other countries, regions, or years.

3.3.1 The Normal random variable

The normal random variable is one of the most important distributions in statistics and probability. It was first introduced by Abraham de Moivre in 1733, who used it to approximate the binomial random variable $Bin(n, p)$ when n is large. This result was later extended to what is known as the **central limit theorem**. This theorem states that the sum of a large number of independent random variables has a distribution approximately normal. We will explore the meaning of this statement after we learn a few facts about the normal random variable.

Definition 3.6 We say X is a normal random variable, or has a normal distribution, if its pdf is

$$f(x) = \frac{1}{\sigma\sqrt{2\pi}} e^{-\frac{(x-\mu)^2}{2\sigma^2}}, \quad -\infty < x < \infty,$$

where μ is any real constant, and σ any positive real constant. X is denoted by $N(\mu, \sigma^2)$.

The expected value and variance of $X \sim N(\mu, \sigma^2)$ are the constants μ and σ^2, respectively. A proof of this statement can be found in introductory textbooks on probability, and it involves using calculus to compute the integrals

$$E[X] = \frac{1}{\sigma\sqrt{2\pi}} \int_{-\infty}^{\infty} x e^{-\frac{(x-\mu)^2}{2\sigma^2}} dx \text{ and } E[X^2] = \frac{1}{\sigma\sqrt{2\pi}} \int_{-\infty}^{\infty} x^2 e^{-\frac{(x-\mu)^2}{2\sigma^2}} dx.$$

Definition 3.7 We say Z has a standard normal distribution if $\mu = 0, \sigma^2 = 1$, and we write $Z \sim N(0, 1)$.

The pdf of a standard normal distribution is $f(x) = \frac{1}{\sqrt{2\pi}} e^{-\frac{x^2}{2}}$, and its cdf is

$$\Phi(x) = \int_{-\infty}^{x} \frac{1}{\sqrt{2\pi}} e^{-\frac{t^2}{2}} \, dt.$$

Theorem 3.4 *If X is $N(\mu, \sigma^2)$, then $Z = \frac{X-\mu}{\sigma}$ is $N(0, 1)$.*

Proof Let's find the cdf of Z :

$$P\{Z \le z\} = P\left\{ \frac{X - \mu}{\sigma} \le z \right\} = P\{X \le \sigma z + \mu\} = \int_{-\infty}^{\sigma z + \mu} \frac{1}{\sigma\sqrt{2\pi}} e^{-\frac{(x-\mu)^2}{2\sigma^2}} \, dx.$$

We now apply change of variables to the integral: $w = (x - \mu)/\sigma$, $dw = dx/\sigma$. Observe that when $x = \sigma z + \mu$, $w = z$, and when $x = -\infty$, $w = -\infty$. Hence the integral becomes

$$P\{Z \le z\} = \int_{-\infty}^{\sigma z + \mu} \frac{1}{\sigma\sqrt{2\pi}} e^{-\frac{(x-\mu)^2}{2\sigma^2}} \, dx = \int_{-\infty}^{z} \frac{1}{\sigma\sqrt{2\pi}} e^{-\frac{w^2}{2}} \sigma \, dw = \int_{-\infty}^{z} \frac{1}{\sqrt{2\pi}} e^{-\frac{w^2}{2}} \, dw,$$

which is simply the cdf of the standard normal distribution. Therefore Z is $N(0, 1)$. $\square$

We can use the above theorem to compute probabilities $P\{a < X < b\}$ when X is $N(\mu, \sigma^2)$, in terms of $\Phi(x)$, the cdf of $N(0, 1)$. To do that, we write

$$P\{a < X < b\} = P\left\{ \frac{a - \mu}{\sigma} < \frac{X - \mu}{\sigma} < \frac{b - \mu}{\sigma} \right\} = \Phi\left(\frac{b - \mu}{\sigma} \right) - \Phi\left(\frac{a - \mu}{\sigma} \right). \quad (3.4)$$

3.3.1.1 Normal distribution in Julia

Julia uses a slightly different notation for the normal distribution. The function Normal(μ, σ) means the normal distribution with mean μ and standard deviation σ. What we should be careful about is when we write, for example, $N(0, 4)$, four is the variance of the random variable, whereas in Julia the same random variable is written as Normal($0, 2$), where two is its standard deviation.

After loading the Distributions and PyPlot packages, we define Z as the standard normal random variable:

```
In [1]: Z=Normal(0,1)
```

```
Out[1]: Normal{Float64}(μ=0.0, σ=1.0)
```

Let's evaluate the pdf and cdf of Z at 0:

```
In [2]: pdf(Z,0)
```

```
Out[2]: 0.3989422804014327
```

```
In [3]: cdf(Z,0)
```

```
Out[3]: 0.5
```

Next we plot the pdf and cdf of Z.

```
In [4]: xval=range(-4,stop=4,length=100)
        y=pdf.(Z,xval)
        plot(xval,y);
```

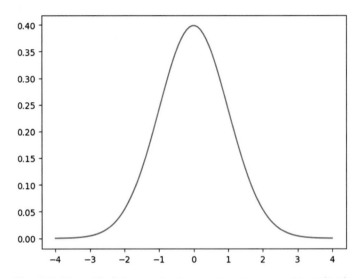

Fig. 3.7: The pdf of the standard normal random variable $N(0, 1)$

```
In [5]: xval=range(-4,stop=4,length=100)
        z=cdf.(Z,xval)
        plot(xval,z);
```

Example 3.4 The actor Chris Hemsworth, better known as the superhero Thor, recounted on a TV show how he stuffed his daughter's shoes with candy bars so that she would meet the height requirement of 40 inches for a Disney ride. Do you wonder how many 5-year-olds get disappointed at Disney World because they are shorter than 40 inches and cannot ride their favorite rides?

It is generally accepted that heights of people roughly follow a normal distribution. For example, the histogram below plots the heights of 746 children[3], and although not a perfect match, the shape of the data resembles the shape of the pdf of a normal random variable (see Figure 3.9).

---

[3] Pearson and Lee's data on the heights of parents and children classified by gender, https://vincentarelbundock.github.io/Rdatasets/datasets.html

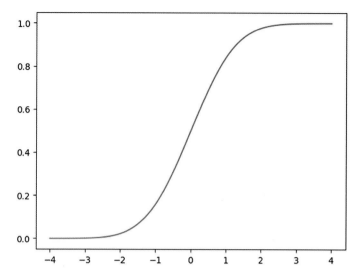

Fig. 3.8: The cdf of the standard normal random variable $N(0, 1)$

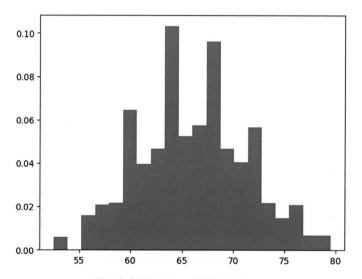

Fig. 3.9: Heights of 746 children

According to a report from the National Center for Health Statistics[4], the average height of 5-year-olds in the U.S. is 44.8 inches, with a standard deviation of 3.1 inches. These values were obtained from a random survey of 205 kids in the age group. Assuming that a normal distribution is a good model for the heights of

[4] Fryar CD, Gu Q, Ogden CL. Anthropometric reference data for children and adults: United States, 200-2010. National Center for Health Statistics. Vital Health Stat 11(252). 2012.

5-year-olds, we can find the proportion of them shorter than 40 inches. Let X be the height of a child, modeled by $N(44.8, 3.1^2)$. Using Eq. (3.4), we obtain

$$P\{X \leq 40\} = P\left\{\frac{X - 44.8}{3.1} \leq \frac{40 - 44.8}{3.1}\right\} = P\{Z \leq -1.5\} = \Phi(-1.5).$$

Recall that earlier we defined Z as a standard normal random variable in Julia. Then we can compute $\Phi(-1.5)$ by using the cdf function of Julia:

In [6]: `cdf(Z,-1.5)`

Out[6]: `0.06680720126885804`

Therefore, about 7% of all 5-year-olds won't be able to enjoy the rides with 40-inch minimum height requirement.

3.3.1.2 Generating random numbers from normal distribution

We generate 10000 random numbers from the standard normal distribution Z and plot their relative frequency histogram using 40 bins in Figure 3.10. Notice how the shape of the histogram follows the shape of the standard normal pdf in Figure 3.7.

In [7]: `hist(rand(Z,10000),40,density=true);`

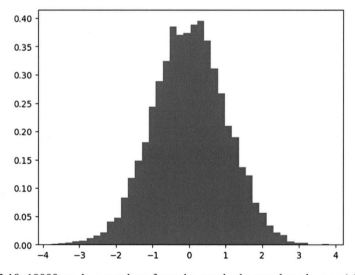

Fig. 3.10: 10000 random numbers from the standard normal random variable Z

3.3.1.3 Central limit theorem

The central limit theorem is one of the most important theorems in probability theory. Let $X_1, X_2, \ldots, X_N$ be independent random variables having the same distribution, and finite variance. Now consider their sample mean

$$S_N = \frac{X_1 + X_2 + \ldots + X_N}{N}$$

which is a random variable as well. We want to know the distribution of S_N.

Let's use Julia to do some experiments. Assume $X_1, X_2, \ldots, X_N$ are independent uniform random variables on $(0, 1)$. We will try to figure out the distribution of S_N empirically in the following way. We will generate N random numbers $x_1, \ldots, x_N$ from $U(0, 1)$, and compute $s_N = \frac{x_1 + x_2 + \ldots + x_N}{N}$. We will then repeat this computation 1000 times, using independent random numbers, to obtain 1000 values for S_N

$$s_N^{(1)}, s_N^{(2)}, \ldots, s_N^{(1000)},$$

and plot the relative frequency histogram of these values.

First we load the packages we need.

```
In [1]: using Distributions
```

```
In [2]: using PyPlot
```

We can generate N random numbers from $U(0, 1)$ using the function **rand(N)**. To add these numbers, we can use the function **sum** and write **sum(rand(N))**. The following code, which is a modification of a code we discussed earlier, takes N as the input and returns the histogram.

```
In [3]: function sMean(N)
            means=zeros(1000)
            for i in 1:1000
                means[i]=sum(rand(N))/N
            end
            hist(means, density=true)
        end
```

```
Out[3]: sMean (generic function with 1 method)
```

```
In [4]: sMean(500)
```

What does the shape of the histogram in Fig. 3.11 remind you of? The normal distribution, isn't it? So it looks like the sample mean random variable S_N may be normally distributed. In our experiment we assumed $X_1, \ldots, X_N$ had the uniform distribution on $(0, 1)$. What would happen if we had picked X_n's from another distribution? Let's try and see.

Next we define X as an exponential random variable with $\theta = 2$:

```
In [5]: X=Exponential(2)
```

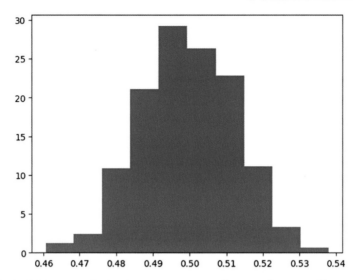

Fig. 3.11: Histogram of S_N when the X_n are uniform random variables on $(0, 1)$

```
Out[5]: Exponential{Float64}(θ=2.0)
```

In the following code, the function **rand(X,N)** generates N random numbers from the distribution X. The rest is similar to the previous code.

```
In [6]: function sMean(N)
            means=zeros(1000)
            for i in 1:1000
                means[i]=sum(rand(X,N))/N
            end
            hist(means, density=true)
        end
```

```
Out[6]: sMean (generic function with 1 method)
```

```
In [7]: sMean(500)
```

Surprise! The distribution of S_N still looks like a normal distribution. And this is precisely what the central limit theorem states! No matter what the distribution of $X_1, \ldots, X_N$ is, as long as the X_n have the same distribution and are independent, $S_N = \frac{X_1+X_2+\ldots+X_N}{N}$ will have an *approximately* normal distribution. The approximation gets better as $N \to \infty$. Here is the precise statement of the central limit theorem:

$$P\left\{\frac{S_N - \mu}{\sigma/\sqrt{N}} \leq a\right\} \to \frac{1}{\sqrt{2\pi}} \int_{-\infty}^{a} e^{-x^2/2} dx \tag{3.5}$$

as $N \to \infty$, where μ and σ^2 are the common expected value and variance of the random variables $X_1, \ldots, X_N$. Note that the limit in Eq. (3.5) is the standard normal

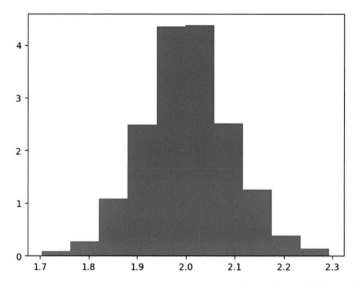

Fig. 3.12: Histogram of S_N when the X_n are exponential random variables with $\theta = 2$

cdf. The expression on the left side of (3.5) is not exactly $P\{S_N \leq a\}$, but the *standardized* version of S_N where we subtract from S_N its mean, μ, and divide the difference by its standard deviation, $\sigma/\sqrt{N}$. In this way the standardized random variables converge to the standard normal distribution. A proof of the central limit theorem can be found in more advanced textbooks on probability theory, such as Billingsley [5].

Let's make one final observation. Earlier we learned from the strong law of large numbers that $S_N = \frac{X_1+X_2+...+X_N}{N}$ converges to $E[X]$ when the X_n are independent and have the same distribution as X. If we inspect Figs. 3.11 and 3.12, we see that the first histogram seems to be symmetric around 0.5, and the second one around 2. This indicates the sample mean of the numbers in the first histogram is about 0.5, and the other 2. Now recall that the expected value of $X \sim U(0, 1)$ is 0.5, and the expected value of $X \sim \exp(\theta = 2)$ is 2. The histograms in Figs. 3.11 and 3.12 illustrate both the strong law of large numbers and the central limit theorem. They show that S_N have a bell-shaped curve, indicating normal distribution, and they are centered at the expected value of the common distribution X where the S_N are obtained from.

3.4 Project 9: Florida Panther

The Florida panther is one of two wild cat species native to Florida, the other being the bobcat. The current territory of Florida panther is southwest Florida. Florida panthers are endangered species, and according to Florida Fish and Wildlife Conservation Commission, there are approximately 120–230 adult panthers in the population as of 2019. Our task is to model the population growth of the panther and assess the likelihood of its extinction.

We will use the following difference equation to model the growth of the Florida panther population:

$$x(n + 1) - x(n) = (b - d)x(n)$$

where $x(n)$ denotes the population at time n, and b, d are the birth and death rates for the panther. The left-hand side of the equation is the change in the population from time n to $n + 1$. The term $b - d$ on the right-hand side measures the net growth rate: for example, if this difference is 0.1, then the model tells us that the panther population will grow by 0.1 times the current population $x(n)$, as we go from time n to time $n + 1$.

In deterministic models, the birth and death rates are taken as constants. A more realistic assumption would be to assume these rates are random quantities. Let's assume the birth rate and the death rate are random variables with a normal distribution:

$$b \sim N(\mu_b, \sigma_b^2)$$
$$d \sim N(\mu_d, \sigma_d^2)$$

The means μ_b, μ_d and variances σ_b^2, σ_d^2 have to be estimated from data.

1. The number of deaths and births of Florida panthers for recent years can be found at https://myfwc.com/wildlifehabitats/wildlife/panther/. Use the available data to estimate the parameters $\mu_b, \mu_d, \sigma_b^2, \sigma_d^2$, using the sample mean and variance of birth and death rate data. Recall that the sample variance of $\{x_1, \ldots, x_n\}$ is

$$\sigma^2 = \frac{1}{n - 1} \sum_{i=1}^{n} (x_i - \bar{x})^2$$

where $\bar{x}$ is the sample mean. The square root of the variance, σ, is the standard deviation. To compute the rate of birth or death, you have to know the current population, for which there are only estimates as cited on the aforementioned web page. For example, if there are 2 births in a given year in a population of 100, then the birth rate is 2%.

You can also use Julia to compute sample means and variances (or, standard deviations). The functions **mean** and **std** compute the sample mean and standard deviation of data. You need a package called **Statistics**:

```
In [1]: using Statistics
```

```
In [2]: list=[12,34,11,4,5,9]
```

```
Out[2]: 6-element Array{Int64,1}:
           12
           34
           11
            4
            5
            9
```

```
In [3]: mean(list)
```

```
Out[3]: 12.5
```

```
In [4]: std(list)
```

```
Out[4]: 11.004544515789828
```

2. Write a Julia code that computes the population $x(n)$ as $n = 1, 2, \ldots, 15$. Try generating a few population trajectories and plotting them.

Recursions can be coded easily in Julia. For example, to code the recursion $x(1) = 1; x(n) = 2x(n-1) + 1$, we write

```
In [1]: function x(n)
            if n==1
                return 1
            else
                2*x(n-1)+1
            end
        end
```

```
Out[1]: x (generic function with 1 method)
```

```
In [2]: [x(n) for n=1:6]
```

```
Out[2]: 6-element Array{Int64,1}:
            1
            3
            7
           15
           31
           63
```

3. The Florida panther will become extinct if the number of breeding animals falls below a critical threshold. Assuming that this threshold corresponds to a population size of 10, estimate the probability that the Florida panther will become extinct in the next 15 years. The Julia function **minimum** will be useful as

```
In [1]: a=[3, 4, 9, 21, 2]

Out[1]: 5-element Array{Int64,1}:
           3
           4
           9
          21
           2

In [2]: minimum(a)

Out[2]: 2
```

3.5 χ^2-distribution and χ^2-test

In the Benford's law Project 2.2, we decided that the distribution of the third digits of the population data was uniform, after we compared its relative frequency histogram visually with that of some random numbers from the uniform distribution. Likewise, earlier in this chapter we claimed data followed exponential or normal distributions based on the similarity of its relative frequency histogram to the pdf of the corresponding random variable. This visual justification is certainly not adequate for any serious work. Pearson, in his seminal paper [20] published in 1900, introduced a quantitative method to decide whether data follows a specific distribution. Interestingly, before the publication of Pearson's paper, the kind of visual qualitative arguments we have used was the only tool available for this purpose. In [20] Pearson criticized his fellow researchers for making judgements based merely on qualitative arguments:

> But the comparison of observation and theory in general amounts to a remark—based on no quantitative criterion—of how well theory and practice really do fit!

The quantitative method Pearson proposed is called the χ^2-test, and it is one of the two main goodness-of-fit tests used today. To learn this test, first we need to learn about a new continuous distribution called the χ^2-distribution.

Definition 3.8 A random variable X is said to have the χ^2-distribution with r degrees of freedom, and written as $X \sim \chi^2(r)$, if its pdf is given by

$$f(x) = \frac{1}{\Gamma(\frac{r}{2})2^{\frac{r}{2}}} x^{\frac{r}{2}-1} e^{-\frac{x}{2}}$$

for $x > 0$, and 0 otherwise.

The function $\Gamma(\cdot)$ that appears in the denominator is called the gamma function and it is defined by

$$\Gamma(t) = \int_0^\infty x^{t-1} e^{-x} dx$$

where $t > 0$. The mean and variance of the χ^2-distribution with r degrees of freedom is

$$E[X] = r, \ Var(X) = 2r.$$

Figure 3.14 plots three χ^2-density functions with 3, 6, and 9 degrees of freedom. Notice how the probability density functions *move right* as r increases: higher degrees of freedom implies the random variable attains larger values with larger probabilities (Fig. 3.13).

Often we need to compute *tail probabilities* like $P\{X > 20\}$. How can we compute $P\{X > 20\}$ if $X \sim \chi^2(9)$? Observe that

$$P\{X > 20\} = 1 - P\{X \le 20\} = 1 - F(20)$$

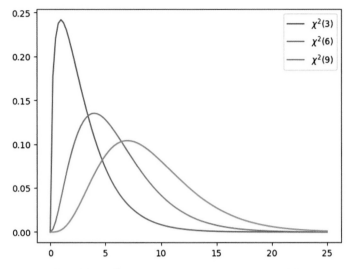

Fig. 3.13: $\chi^2(r)$ density functions for $r = 3, 6, 9$

where F is the cdf of X. We can compute this using Julia. After we load the package
Distributions, we define $X \sim \chi^2(9)$ by simply typing, $X = \text{Chisq}(9)$:

```
In [1]: X=Chisq(9)
        1-cdf.(X,20)
```

```
Out[1]: 0.01791240452984333
```

Therefore the area under the density function of $\chi^2(9)$ to the right of 20 is about
0.018.

Now let's discuss the χ^2-test. Here is the setup: we have a random experiment
that has k mutually exclusive and exhaustive outcomes, say $A_1, \ldots, A_k$. And we have
a hypothesis, called the *null hypothesis*, which claims the probability of the outcome
A_i is p_i. We want to test this hypothesis, whether the probability of A_i is really p_i.

To test the null hypothesis, we repeat the random experiment, say n times, and
count how many times each outcome appears. Let C_i be the number of times outcome
A_i appears. If the null hypothesis is correct, we would expect A_i to appear about np_i
times. In other words, we expect C_i to be close to np_i. If C_i and np_i are very far
from each other, then we might suspect the null hypothesis is wrong. The larger the
difference between C_i and np_i is, the less plausible the null hypothesis is correct.
Consider the following expression based on these differences:

$$Q_{k-1} = \sum_{i=1}^{k} \frac{(C_i - np_i)^2}{np_i}. \tag{3.6}$$

The quantity Q_{k-1} is a real number; but every time we repeat the random experiment
n times, we will obtain different values for C_i, hence a different value for Q_{k-1}.

Therefore Q_{k-1} is a random variable, which outputs a positive real number through the expression in Eq. (3.6), every time the random experiment is repeated n times.

Here is an example: suppose the random experiment is rolling a die. Assume the null hypothesis claims the die is fair, and thus $P(i) = 1/6$ where $i = 1, 2, \ldots, 6$. We roll the die 60 times and obtain the following results:
Then, Q_5 is

$$\text{Outcome } 1 \ 2 \ 3 \ 4 \ 5 \ 6$$
$$C_i \quad 8 \ 7 \ 12 \ 10 \ 9 \ 14$$

$$Q_5 = \frac{(8 - 10)^2}{10} + \frac{(7 - 10)^2}{10} + \frac{(12 - 10)^2}{10} + \frac{(10 - 10)^2}{10} + \frac{(9 - 10)^2}{10} + \frac{(14 - 10)^2}{10}$$
$$= 3.4.$$

We then roll the die again, 60 times, and obtain

$$\text{Outcome } 1 \ 2 \ 3 \ 4 \ 5 \ 6$$
$$C_i \quad 12 \ 2 \ 4 \ 9 \ 16 \ 17$$

The new value for Q_5 is

$$Q_5 = \frac{(12 - 10)^2}{10} + \frac{(2 - 10)^2}{10} + \frac{(4 - 10)^2}{10} + \frac{(9 - 10)^2}{10} + \frac{(16 - 10)^2}{10} + \frac{(17 - 10)^2}{10}$$
$$= 19.0.$$

We established that Q_{k-1} is a random variable. A natural question is, what is the distribution of Q_{k-1}? Pearson [20] showed that Q_{k-1} has approximately a χ^2-distribution with degrees of freedom equal to $k - 1$, and this approximation gets better with larger n. This fact is all we need to design a quantitative test to check if data follows the probabilities specified by the null hypothesis. Here is how we argue: consider the previous die example where we had $Q_5 = 19$. Recall that Q_5 follows the χ^2-distribution with 5 degrees of freedom, whose density function is plotted in Fig 3.14:

How likely is it that a random variable with the above density function attains a value like 19? This is not a very good question, for the probability that a continuous random variable attains a specific value is always 0. Let's ask instead, how likely is it that a random variable with the above density function takes a value greater than 19? This is the tail probability $P\{X > 19\}$ and earlier we discussed how to compute it using the cdf of X. Using Julia, we compute

```
In [1]: X=Chisq(5)
        1-cdf.(X,19)
```

```
Out[1]: 0.001922136820942999
```

The probability $P\{X > 19\} \approx 0.0019$ is a *very small number*. The χ^2-distribution with 5 degrees of freedom gives a value of 19 or more only twice in a thousand. This makes us suspicious.

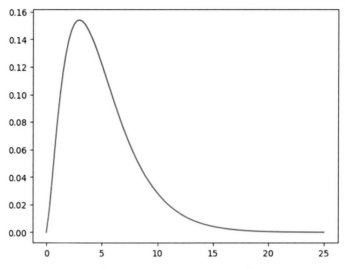

Fig. 3.14: Density function for $\chi^2(5)$

Let's recap where we are: we want to test if the null hypothesis is correct or not. If it were correct, then Q_5 would follow the χ^2-distribution with 5 degrees of freedom. But we observed a value of 19 in our die experiment, which is a very rare event (two in a thousand) if the null hypothesis were correct. Then our conclusion is the null hypothesis is *not correct*. We reject the null hypothesis, and conclude the die is not fair.

Here is a summary of how we use the χ^2-test:

- There is a random experiment with outcomes $A_1, \ldots, A_k$, and a null hypothesis, which claims $P(A_i) = p_i$. We want to test if the null hypothesis is correct.
- Repeat the random experiment n times, and count how many times each outcome appears. Let C_i be the number of times outcome A_i appears. Compute

$$Q_{k-1} = \sum_{i=1}^{k} \frac{(C_i - np_i)^2}{np_i}.$$

- If Q_{k-1} is too large (or, too small), reject the null hypothesis.

One loose point in the above description is the phrase *too large* in the last step. What exactly does that mean? In the die example, we decided 19 was too large, because the *tail probability* that the random variable gave a value larger than 19 was 0.0019. A rule of thumb is to reject the null hypothesis whenever the tail probability is less than 5%. This is a threshold we set: anything less than that is too small and the null hypothesis is suspect. Sometimes researchers use different thresholds matching their preference for what a rare event should be.

3.6 Project 10: Can humans generate random numbers?

It looks like the jury is still out on this question, at least in the medical community. Persaud [21] claims humans can generate random numbers, and Figurska et al. [7] claim they cannot.

In this project we will generate *random* digits 0 through 9, and then test how truly random these digits are. We will compare our performance with that of a machine: the random generator of Julia.

1. Generate 500 digits 0 through 9 at random[5] input the digits in Julia as an array, like

```
In [1]: digs=[1, 2, 0, 9, 3, 3, 6, 2]

Out[1]: 8-element Array{Int64,1}:
          1
          2
          0
          9
          3
          3
          6
          2
```

   a. If the numbers in **digs** are truly random, then the relative frequency of each digit (0 through 9) should be 1/10. Test this hypothesis using the χ^2-test. As you write a code for the χ^2-statistic, you may find the **count** function useful. Here is how the **count** function can be used to count the number of 2's in the array **digs**:

```
In [2]: count(n->n==2,digs)
Out[2]: 2
```

   b. Next we will test the distribution of pairs of digits. If the numbers in **digs** are truly random, then the relative frequency of any consecutive pair of digits should be 1/100. Think of the random experiment as generating pairs of digits. There are 11 possible outcomes: 00, 11, 22, 33, 44, 55, 66, 77, 88, 99, and "all others". (We are grouping pairs that do not repeat in one category.) The probabilities of the outcomes are 1/100 for the equal-pair outcomes, and 9/10 for the last outcome "all others". Find the frequency of each outcome in your data and carry out the χ^2-test and check the randomness of the data. The following functions will be helpful:

```
In [3]: pairs=[(digs[2*i-1],digs[2*i]) for i in 1:4]
Out[3]: 4-element Array{Tuple{Int64,Int64},1}:
          (1, 2)
          (0, 9)
```

---

[5] YSP students at Florida State University generate the digits collectively. In our case, each student generated 25 random digits,

```
                    (3, 3)
                    (6, 2)
In [4]: count(n->n==(3,3),pairs)
Out[4]: 1
```

2. Apply the χ^2-test as described in parts (a) and (b) to the random numbers generated by Julia.

3. What are your conclusions? Do you think humans can generate random numbers? How about Julia?

Chapter 4
Markov Chains

4.1 Introduction to Markov chains

Until now, we mostly dealt with a single random variable at a time and used it to model a random quantity, such as the waiting time for a customer to enter a store or the number of deaths due to horse kicks. Occasionally the problem gave rise to a sequence of independent random variables like in the coupon collector problem. We studied tools that help us with the analysis of single random variables and sequences of independent random variables. However, in applications we frequently encounter sequences of random variables that are *not* independent. Andrey Markov was one of the firsts who gave a systematic analysis of such sequences in the early 1900s. Today we call these sequences **Markov chains**.

Example 4.1 (Wanderings of an iPhone zombie)
    Consider a typical FSU student doing things on his phone while walking on a straight path. At each step, the student goes to his right or left, with equal probability $1/2$. The starting point is taken as the origin. Let X_n be the coordinate of the student after n steps. Note that

$$X_0 = 0;$$

$$X_1 = \begin{cases} 1, & \text{with probability } 1/2 \\ -1, & \text{with probability } 1/2 \end{cases};$$

$$X_2 = \begin{cases} 2, & \text{with probability } 1/4 \\ 0, & \text{with probability } 1/2 \\ -2, & \text{with probability } 1/4 \end{cases}.$$

Clearly, keeping track of the student and the probabilities can get complicated as he takes more steps. However, it is always simple to figure out the next location X_{n+1}, from the current location X_n. If $X_n = i$, then

$$X_{n+1} = \begin{cases} i + 1, & \text{with probability } 1/2 \\ i - 1, & \text{with probability } 1/2 \end{cases}.$$

© The Editor(s) (if applicable) and The Author(s), under exclusive
license to Springer Nature Switzerland AG 2020
G. Ökten, *Probability and Simulation*, Springer Undergraduate Texts
in Mathematics and Technology, https://doi.org/10.1007/978-3-030-56070-6_4

Note that the probability of X_{n+1} is completely determined from X_n: we do not need to know about the location prior to step n.

Definition 4.1 (Markov chain) We say that the sequence $\{X_0, X_1, \dots\}$ of random variables is a Markov chain if the probability that $\{X_{n+1} = j\}$ depends only on X_n.

We define the **transition probabilities** of a Markov chain as

$$p_{ij} = P\{X_{n+1} = j \mid X_n = i\}.$$

We also define the **state space** of a Markov chain as the set of all possible values the random variables X_n can take. Notice that p_{ij} does not depend on the index n; such Markov chains are called **homogeneous**. In the iPhone zombie example, the state space is the set of all integers $\{\dots, -2, -1, 0, 1, 2, \dots\}$.

In 1913 Markov wrote *"An example of statistical investigation of the text Eugene Onegin concerning the connection of samples in chains"* [16] where he analyzed the letter transitions in the novel Eugene Onegin by the great writer Pushkin. Markov examined 20,000 letters from the novel, and calculated transition probabilities such as the probability that a vowel follows another vowel, and a vowel follows a consonant. Today Markov chains are widely used in many fields including physical and social sciences, and economics and finance.

Example 4.2 (Weather forecasting) Suppose that the chance of rain tomorrow depends on whether it rains today. If it rains today, it will rain tomorrow with probability α. If it does not rain today, it will rain tomorrow with probability β. Let

$$X_n = \begin{cases} 0, & \text{if it rains on day } n \\ 1, & \text{if it does not rain on day } n \end{cases}.$$

where $n = 0$ corresponds to today. Then $\{X_0, X_1, \dots\}$ is a Markov chain with state space $\{0, 1\}$. It is a Markov chain because the probability of the events $\{X_{n+1} = 0\}$ and $\{X_{n+1} = 1\}$ depends only on the value of X_n, as can be seen from the transition probabilities below:

$$p_{00} = P\{X_{n+1} = 0 \mid X_n = 0\} = \alpha$$
$$p_{10} = P\{X_{n+1} = 0 \mid X_n = 1\} = \beta$$
$$p_{01} = P\{X_{n+1} = 1 \mid X_n = 0\} = 1 - \alpha$$
$$p_{11} = P\{X_{n+1} = 1 \mid X_n = 1\} = 1 - \beta$$

These probabilities are usually written in matrix form, called the **transition matrix**:

$$P = (p_{ij}) = \begin{bmatrix} p_{00} & p_{01} \\ p_{10} & p_{11} \end{bmatrix} = \begin{bmatrix} \alpha & 1 - \alpha \\ \beta & 1 - \beta \end{bmatrix}$$

The Markov chain is homogeneous because the transition probabilities do not depend on the index n, the time. Rainy weather follows rainy weather with a fixed probability α, independent of what day it is.

A graphical representation of the states and transition probabilities is given by the **state transition diagram**.

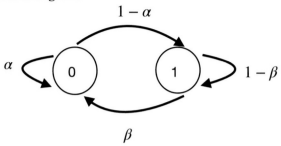

Using the simple Markov chain model for weather forecasting, how can we compute the probability that it will rain the next day if it rains today? Or it will rain 10 days later if it rains today? These probabilities are called **higher transition probabilities**.

4.1.1 Higher transition probabilities

Let $p_{ij}(n)$ be the probability that the Markov chain that is currently in state i will be in state j after n transitions

$$p_{ij}(n) = P\{X_{n+m} = j \mid X_m = i\}, n \geq 0, \tag{4.1}$$

where m can be any positive integer; the conditional probability does not depend on the value of m from the homogeneity of the Markov chain. The probability $p_{ij}(n)$ is called the n-step transition probability.

Going back to the weather forecasting example, let's compute

$$p_{00}(2) = P\{X_2 = 0 \mid X_0 = 0\}.$$

Since the event $\{X_2 = 0\}$ can be written as the union of two disjoint sets $\{X_2 = 0, X_1 = 0\}$ and $\{X_2 = 0, X_1 = 1\}$, we have

$$P\{X_2 = 0\} = P\{X_2 = 0, X_1 = 0\} + P\{X_2 = 0, X_1 = 1\}.$$

The above identity also holds true if the probabilities are replaced by conditional probabilities:

$$P\{X_2 = 0 \mid X_0 = 0\} = P\{X_2 = 0, X_1 = 0 \mid X_0 = 0\} + P\{X_2 = 0, X_1 = 1 \mid X_0 = 0\}. \tag{4.2}$$

Let's compute the first probability on the right-hand side. Using the definition of conditional probability repeatedly, we have

$$P\{X_2 = 0, X_1 = 0 \mid X_0 = 0\} = \frac{P\{X_2 = 0, X_1 = 0, X_0 = 0\}}{P\{X_0 = 0\}}$$

$$= \frac{P\{X_2 = 0 \mid X_1 = 0, X_0 = 0\}P\{X_1 = 0, X_0 = 0\}}{P\{X_0 = 0\}}$$

$$= \frac{P\{X_2 = 0 \mid X_1 = 0, X_0 = 0\}P\{X_1 = 0 \mid X_0 = 0\}P\{X_0 = 0\}}{P\{X_0 = 0\}}.$$

Since X_n is a Markov chain, the conditional probability $P\{X_2 = 0 \mid X_1 = 0, X_0 = 0\}$ is the same as $P\{X_2 = 0 \mid X_1 = 0\} = p_{00} = \alpha$—what happens on day two only depends on day 1, not day 0. Since $P\{X_1 = 0 \mid X_0 = 0\} = p_{00} = \alpha$ as well, and after dividing out by $P\{X_0 = 0\}$, we get

$$P\{X_2 = 0, X_1 = 0 \mid X_0 = 0\} = \alpha^2. \qquad (4.3)$$

In a similar way, we can show that

$$P\{X_2 = 0, X_1 = 1 \mid X_0 = 0\} = (1 - \alpha)\beta. \qquad (4.4)$$

Substituting these two probabilities given by Eqs. (4.3) and (4.4), into Eq. (4.2), we conclude

$$p_{00}(2) = P\{X_2 = 0 \mid X_0 = 0\} = \alpha^2 + (1 - \alpha)\beta.$$

These lengthy calculations can be done very quickly using a probability tree diagram.

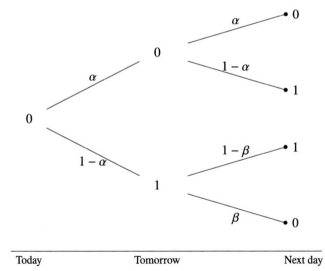

| Today | Tomorrow | Next day |

Rain on day 2 (next day) is possible through the paths $0 \to 0 \to 0$ and $0 \to 1 \to 0$ in the tree diagram above. The probability of each path is computed by multiplying the probabilities on each branch, and then we add the probabilities of different paths. This gives α^2 for the first path, $(1 - \alpha)\beta$ for the second path, and their sum is the probability we obtained for $p_{00}(2)$ earlier. Similarly, we can compute $p_{01}(2)$ following the branches of the probability tree diagram:

$$p_{01}(2) = P\{X_2 = 1 \mid X_0 = 0\} = \alpha(1 - \alpha) + (1 - \alpha)(1 - \beta).$$

Although the tree diagram makes it easier to compute the conditional probabilities, it has its limitations. Imagine extending the diagram to 10 days to compute $p_{01}(10)$: we would need a much larger paper and a lot of patience! There is, however, a much simpler way. Let's compute the square of the transition matrix:

$$P^2 = \begin{bmatrix} \alpha & 1-\alpha \\ \beta & 1-\beta \end{bmatrix} \times \begin{bmatrix} \alpha & 1-\alpha \\ \beta & 1-\beta \end{bmatrix} = \begin{bmatrix} \underbrace{\alpha^2 + (1-\alpha)\beta}_{p_{00}(2)} & \underbrace{\alpha(1-\alpha) + (1-\alpha)(1-\beta)}_{p_{01}(2)} \\ \beta\alpha + (1-\beta)\beta & \beta(1-\alpha) + (1-\beta)^2 \end{bmatrix}$$

Notice that $p_{00}(2)$ is the $(0,0)$th entry of the matrix P^2 (labeling the rows and columns as 0,1, matching the way Markov chain states are numbered), and $p_{01}(2)$ is the $(0,1)$th entry of P^2. It turns out that this surprising observation is always true for any Markov chain with finitely many states, and for any higher transition probability! The higher transition probabilities can be computed by simply taking matrix powers of P, the transition matrix of the Markov chain. The next theorem states this fact, and a proof can be found in most probability textbooks.

Theorem 4.1 *Let P be the transition matrix of a Markov chain. Then, the ijth entry of P^n, the nth power of the matrix P, is $p_{ij}(n)$.*

Example 4.3 Let $\alpha = 0.7, \beta = 0.4$ in the weather forecasting example. Find the probability that it will rain 4 days from today given that it is raining today.

The transition matrix is

$$P = \begin{bmatrix} \alpha & 1-\alpha \\ \beta & 1-\beta \end{bmatrix} = \begin{bmatrix} 0.7 & 0.3 \\ 0.4 & 0.6 \end{bmatrix}.$$

We need to compute P^4. We get help from Julia:

```
In [1]: P=[.7 .3;.4 .6]

Out[1]: 2×2 Array{Float64,2}:
        0.7  0.3
        0.4  0.6

In [2]: P^4

Out[2]: 2×2 Array{Float64,2}:
        0.5749  0.4251
        0.5668  0.4332
```

Therefore the desired probability is $p_{00}(4) = 0.5749$.

4.2 Project 11: Analyzing a die game with Markov chains

This game was introduced by Lawrance [12]. Consider a die with one face labeled "I won!", two faces labeled "Roll it again", and three faces labeled "Pass it to the opponent". Two players play this game by simply rolling the die and following the instructions on the uppermost face of the die. The game ends with the winner rolling the face "I won!". We want to know the probability that the player who starts the game, that is the player who rolls first, wins the game.

1. Play the game repeatedly to estimate the probability in question.
2. We can analyze this game using Markov chains. The states of the chain correspond to the four different stages the game can be at any roll:

   - State 1': Player 1 wins.
   - State 1: Player 1 rolls the die.
   - State 2: Player 2 rolls the die.
   - State 2': Player 2 wins.

   A part of the state transition diagram for the Markov chain is given in Figure 4.1. Complete the diagram by finding the rest of the transitions and their probabilities. Note that when the Markov chain reaches the state 1' or 2', it stays at those states: this is how we interpret a player's winning the game.
3. Find the transition probability matrix P. How do you interpret the n-step transition probability $p_{11'}(n)$?
4. Use Julia to compute P^{50} and P^{100}. Does it look like the matrix P^n is converging to a limiting matrix P^∞?
5. Using P^∞, find the probability that the player who starts the game, wins the game.

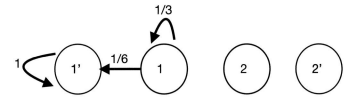

Fig. 4.1: State transition diagram

4.3 State vectors and limiting probabilities

Monopoly is a popular board game with a history going back to 1903. Players roll dice to get around on the squares of the game board, and in some squares a player can buy a property, build a house or a hotel on it, and collect rent from the opponents who visit them. There are 40 squares, and all players start on the GO square. The game can be modeled using a Markov chain with some simplifying assumptions (see Stewart [23]), where the 40 squares are the states of the Markov chain, and there are 40^2 possible transition probabilities that need to be calculated.

Do you wonder if there are some squares a player visits more often than others in Monopoly? If you play the game for a long time, which squares will you end up visiting more often? These questions can be answered using some tools we will learn in this section. In the game of Monopoly as modified by Stewart, it turns out that in the long run players spend most of their time in the Jail, with a likelihood of 5.89%! The next most visited square is the Illinois Ave.

To answer questions regarding the long-term behavior of a Markov chain, we start with introducing the **state vector** which keeps track of the probability of the Markov chain being in each state at any time.

Definition 4.2 (State vector) The state vector for a Markov chain $\{X_0, X_1, \ldots\}$ is the row vector

$$\pi_n = [\pi_n(1), \pi_n(2), \ldots, \pi_n(N)]$$

where $\pi_n(j) = P\{X_n = j\}$ and the states of the Markov chain are $\{1, 2, \ldots, N\}$.

The state vector π_n can be easily computed from its initial value π_0 and the transition matrix P of the Markov chain. Using the law of total probability, we can write $\pi_n(j)$ as

$$\begin{aligned}
\pi_n(j) &= P\{X_n = j\} \\
&= P\{X_n = j \mid X_0 = 1\}P\{X_0 = 1\} + \ldots + P\{X_n = j \mid X_0 = N\}P\{X_0 = N\} \\
&= p_{1j}(n)\pi_0(1) + \ldots + p_{Nj}(n)\pi_0(N). \tag{4.5}
\end{aligned}$$

Observe that the left-hand side of Eq. (4.5) is the jth component of the vector π_n, and the right-hand side of Eq. (4.5) is the jth component of the matrix product

$\pi_0 \times P^n$ (recall that the ijth entry of P^n is $p_{ij}(n)$ from Theorem 4.1):

$$\underbrace{\left[\pi_0(1)\ \pi_0(2)\ \ldots\ \pi_0(N)\right]}_{\pi_0} \times \underbrace{\begin{bmatrix} p_{11}(n) & \ldots & p_{1j}(n) & \ldots & p_{1N}(n) \\ p_{21}(n) & \ldots & p_{2j}(n) & \ldots & p_{2N}(n) \\ \vdots & & \vdots & & \vdots \\ p_{N1}(n) & \ldots & p_{Nj}(n) & \ldots & p_{NN}(n) \end{bmatrix}}_{P^n}$$

Therefore Eq. (4.5) implies

$$\pi_n = \pi_0 P^n.$$

Another useful identity is

$$\pi_{n+1} = \pi_n P.$$

To prove this, observe

$$\pi_{n+1}(j) = P\{X_{n+1} = j\} = \sum_{i=1}^{N} P\{X_{n+1} = j \mid X_n = i\} P\{X_n = i\} = \sum_{i=1}^{N} \pi_n(i) p_{ij},$$

and that the left-hand side is the jth component of the vector π_{n+1}, and the right-hand side is the jth component of the vector matrix product $\pi_n P$.

The following lemma summarizes our findings.

Lemma 4.1 *Let π_n be the state vector, and P be the transition matrix of a Markov chain. Then, we have $\pi_{n+1} = \pi_n P$ and $\pi_n = \pi_0 P^n$.*

4.3.1 Limiting behavior of the state vector

The two largest libraries at FSU are Dirac and Strozier. Assume Dirac has 30% of all the books, and Strozier has the remaining 70% , at the beginning of a new semester. When a student checks out a book, she can return it to any library. A survey of students during the first 2 weeks of the semester shows that 80% of students who check out a book from Dirac return the book to Dirac, and the rest is returned to Strozier. Students who check out from Strozier return their books to Strozier at the rate of 60%. We need to find how many books will end up in each library as the semester unwinds; Dirac has a smaller capacity for storing books, and if too many books are returned there, we need to transport books periodically from Dirac to Strozier.

We will model this problem using Markov chains. The states are D (Dirac) and S (Strozier). The initial proportions of books are given by $\pi_0 = [0.3, 0.7]$. The transition probabilities are

$$p_{DD} = 0.8, p_{DS} = 0.2, p_{SS} = 0.6, p_{SD} = 0.4,$$

and the probability transition matrix is

```
In [1]: P=[0.8 0.2; 0.4 0.6]
```

```
Out[1]: 2×2 Array{Float64,2}:
        0.8  0.2
        0.4  0.6
```

In this example time is measured in terms of loan cycles. Let's simplify the loan policy and assume students must return their books exactly 1 week after they borrow them. Also assume all of the books are in circulation all the time. After n cycles, the proportion of books in Dirac and Strozier will be given by the state vector

$$\pi_n = [\pi_n(D), \pi_n(S)],$$

and from Lemma 4.1, $\pi_n = \pi_0 P^n$, which is coded in Julia as

```
In [2]: pi(n)=[.3;.7]'*P^n
```

```
Out[2]: pi (generic function with 1 method)
```

Here is the state vector at $n = 5$

```
In [3]: pi(5)
```

```
Out[3]: 1×2 LinearAlgebra.Adjoint{Float64,Array{Float64,1}}:
        0.662912  0.337088
```

and the state vector at $n = 20$:

```
In [4]: pi(20)
```

```
Out[4]: 1×2 LinearAlgebra.Adjoint{Float64,Array{Float64,1}}:
        0.666667  0.333333
```

It looks like the state vector π_n is converging to the vector $\pi = [0.67, 0.33]$ as $n \to \infty$. In other words for large n, $\pi_n(D) \approx 0.67$ and $\pi_n(S) \approx 0.33$, which means 67% of the books end up in Dirac, and 33% in Strozier, and thus there will have to be periodic transportation of books from Dirac to Strozier.

4.3.1.1 Limiting behavior of the transition matrix

In Section 4.2, we computed higher powers of the transition matrix P to investigate whether the transition probabilities converged to a limit. Let's do the same analysis for the library example. Here are P^{10} and P^{20}:

```
In [5]: P^10

Out[5]: 2×2 Array{Float64,2}:
        0.666702  0.333298
        0.666597  0.333403

In [6]: P^20

Out[6]: 2×2 Array{Float64,2}:
        0.666667  0.333333
        0.666667  0.333333
```

It looks like the transition matrix P is converging to the following matrix:

$$P^\infty = \begin{bmatrix} 0.67 & 0.33 \\ 0.67 & 0.33 \end{bmatrix}.$$

What is interesting about P^∞ is that each row is equal to $\pi = [0.67, 0.33]$, the limit of the state vector π_n we discussed earlier. Here is a summary of our observations from the library example:

- The state vector π_n seems to converge to a vector: $\pi_n \to \pi$ as $n \to \infty$. This limit vector π is called the **steady-state vector**.
- The transition probability matrix P^n converges to a matrix P^∞ where each row is the steady-state vector π.

Do these observations generalize to any Markov chain? Let's start with the first one: does $\{\pi_n\}$ always have a limit as $n \to \infty$? The answer is no, as the following example shows.

Example 4.4 Consider a Markov chain $\{X_0, X_1, \ldots\}$ with state space $\{0, 1\}$, and transition matrix

$$P = \begin{bmatrix} 0 & 1 \\ 1 & 0 \end{bmatrix}.$$

Assume that $X_0 = 1$, that is, the chain starts in state 1. Observe that

$$\pi_0 = [\pi_0(0), \pi_0(1)] = [P\{X_0 = 0\}, P\{X_0 = 1\}] = [0, 1].$$

Using Lemma 4.1, let's compute the first few terms of the sequence $\{\pi_0, \pi_1, \pi_2, \ldots\}$:

$$\pi_1 = \pi_0 P = [0, 1] \begin{bmatrix} 0 & 1 \\ 1 & 0 \end{bmatrix} = [1, 0]$$

$$\pi_2 = \pi_1 P = [1, 0] \begin{bmatrix} 0 & 1 \\ 1 & 0 \end{bmatrix} = [0, 1]$$

$$\pi_3 = \pi_2 P = [0, 1] \begin{bmatrix} 0 & 1 \\ 1 & 0 \end{bmatrix} = [1, 0].$$

The pattern should be clear: the sequence $\{\pi_0, \pi_1, \pi_2, \ldots\}$ alternates between the vectors $[0, 1]$ and $[1, 0]$, and thus the sequence does not have a limit.

4.3.1.2 Finding the steady-state vector when it exists

From Example 4.4 we learned that not all Markov chains have a steady-state vector. Mathematicians have discovered which Markov chains possess steady-state vectors, but we will not get into that topic. Instead, we will discuss how to find the steady-state vector π, when it actually exists. Earlier we guessed the value of π by computing large values of π_n and guessing what the limit is. However, guesswork has its own limitations, and we want a foolproof way of finding the limit!

Suppose the limit exists, that is, $\lim_{n \to \infty} \pi_n = \pi$. From Lemma 4.1 we have

$$\pi_{n+1} = \pi_n P.$$

Taking the limit of both sides as $n \to \infty$ we get

$$\pi = \pi P.$$

Also observe that $\sum_{i=1}^{N} \pi_n(i) = 1$ for any n, since the components of the vector π_n are the probabilities that the Markov chain is in states $1, 2, \ldots N$, at time n. Taking the limit as $n \to \infty$ gives $\sum_{i=1}^{N} \pi(i) = 1$.

We will find the steady-state vector π by solving the equations $\pi = \pi P$ and $\sum_{i=1}^{N} \pi(i) = 1$ for π. Let's see how to solve these equations for the library example we discussed earlier. The steady-state vector in the library example was $\pi = [\pi(D), \pi(S)]$. The equations we need to solve are

$$\pi(D) + \pi(S) = 1$$

$$\begin{bmatrix} \pi(D) & \pi(S) \end{bmatrix} = \begin{bmatrix} \pi(D) & \pi(S) \end{bmatrix} \times \begin{bmatrix} 0.8 & 0.2 \\ 0.4 & 0.6 \end{bmatrix}$$

Multiplying the vector and matrix, we get the following equations:

$$\pi(D) + \pi(S) = 1$$
$$0.8\pi(D) + 0.4\pi(S) = \pi(D)$$
$$0.2\pi(D) + 0.6\pi(S) = \pi(S)$$

There are 2 unknowns and 3 equations. It turns out that one of the last two equations is redundant. So take the first equation, and any one from the last two, and solve the resulting system. Here we take the first two equations, and rewrite them as

$$\pi(D) + \pi(S) = 1$$
$$-0.2\pi(D) + 0.4\pi(S) = 0$$

This is a simple system of equations we can solve by hand, but let's practice solving it with Julia: the system of equations can be written as $Ax = c$, where A is the coefficient matrix, and c is the constant vector:

$$\underbrace{\begin{bmatrix} 1 & 1 \\ -0.2 & 0.4 \end{bmatrix}}_{A} \underbrace{\begin{bmatrix} \pi(D) \\ \pi(S) \end{bmatrix}}_{x} = \underbrace{\begin{bmatrix} 1 \\ 0 \end{bmatrix}}_{c}$$

To solve this in Julia for x, we simply type "A backslash c":

```
In [7]: A=[1 1 ;-0.2 0.4]
        c=[1; 0];
```

```
In [8]: A\c
```

```
Out[8]: 2-element Array{Float64,1}:
        0.6666666666666667
        0.3333333333333333
```

Therefore $\pi(D) \approx 0.67$, $\pi(S) \approx 0.33$, and $\pi = [0.67, 0.33]$. Notice that this is the same vector we got before when we estimated π by estimating the limit of π_n.

4.3.2 Parrando's paradox

Consider a game of chance where you either win \$1, or lose \$1. For example, imagine a game where you roll a fair die, and win \$1 if the outcome is 5, 6, and lose \$1 if the outcome is 1, 2, 3, 4. The probability of losing is 2/3 which is more than a half, and clearly you will be losing money as you play this game repeatedly. We call such a game, that is a game where the probability of losing is more than one half, a *losing game*. Likewise, if the probability of winning is more than 1/2, we call the game a *winning game*.

Now imagine there are two losing games, Game A and Game B, and you repeatedly play these games according to some strategy. The strategy tells you when to play Game A, and when to play Game B. Let's call this a "hybrid" game, Game C. Here is the question: could there be a strategy under which Game C is a winning game, even though the individual games it is obtained from are losing games? The answer is yes! This counterintuitive result is known as Parrando's paradox, discovered by Juan Manuel Rodriguez Parrondo in 1997.

As we play games repeatedly, we will keep track of our *capital*. We may start playing with an initial capital, and then update how much money we have as we win and lose. We now describe how to construct a winning game from two losing games.

4.3.2.1 Game A

Flip a biased coin where $P(H) = \frac{1}{2} - \epsilon$. Here H refers to the outcome "heads", and ϵ is some small positive number. If the outcome is H, we win \$1. Otherwise we lose \$1. Clearly, Game A is a losing game since $P(\text{win } \$1) = 1/2 - \epsilon < 1/2$ for ϵ is a positive number.

4.3.2.2 Game B

This game is a little more involved. Here we have two biased coins with the following probabilities:

- Coin 1: $P(H) = \frac{1}{10} - \epsilon$
- Coin 2: $P(H) = \frac{3}{4} - \epsilon$,

where, as before, ϵ is some small positive number.

Here is how to play Game B. If our current capital is a multiple of 3, flip Coin 1, otherwise flip Coin 2. In each case, "heads" wins \$1, and "tails" loses \$1.

Now we will show that Game B is a losing game, using the tools we learned in this chapter. Let X_n denote the remainder when our capital is divided by three at the nth time we play the game. Clearly, X_n takes the values 0, 1, or 2. Moreover, X_n is a Markov chain with the following transition matrix

$$P = \begin{bmatrix} p_{00} & p_{01} & p_{02} \\ p_{10} & p_{11} & p_{12} \\ p_{20} & p_{21} & p_{22} \end{bmatrix} = \begin{bmatrix} 0 & \frac{1}{10} - \epsilon & \frac{9}{10} + \epsilon \\ \frac{1}{4} + \epsilon & 0 & \frac{3}{4} - \epsilon \\ \frac{3}{4} - \epsilon & \frac{1}{4} + \epsilon & 0 \end{bmatrix}.$$

Let's verify some of these probabilities. Consider $p_{11} = P\{X_{n+1} = 1 | X_n = 1\}$. At time n, $X_n = 1$, and we play the next round. For X_{n+1} to be equal to 1, one of the following two cases must happen:

1. We won no money at time $n + 1$;
2. We won a multiple of \$3, say $3k$ where k is some positive integer. Our capital has increased by $3k$, but since the remainder when $3k$ is divided by 3 is zero, $X_{n+1} = X_n = 1$. [1]

However none of these cases can really happen. Each time we play, we either lose or win \$1. Winning no money, or multiples of \$3, is not possible. Therefore $p_{11} = 0$.

---

[1] For readers who are familiar with modular arithmetic, this can be simply written as $X_{n+1} = 1 + 3k \equiv 1 \pmod 3$.

Let's verify $p_{21} = 1/4 + \epsilon$. We have $p_{21} = P\{X_{n+1} = 1 | X_n = 2\}$. If our capital is \$2 at time n, according to the rules, we flip Coin 2. To make a transition to $X_{n+1} = 1$, we have to lose \$1. Losing \$1 with Coin 2 happens with probability $1 - (3/4 - \epsilon) = 1/4 + \epsilon$.

How about $p_{02} = P\{X_{n+1} = 2 | X_n = 0\}$? At time n, $X_n = 0$, which means our capital is a multiple of 3, say $3k$ for some integer k. How can we have $X_{n+1} = 2$? If we win \$1, the capital becomes $3k + 1$, which gives a remainder of 1 when divided by 3. If we lose \$1, the capital becomes $3k - 1$, which gives a remainder of 2 when divided by 3. So transition from $X_n = 0$ to $X_{n+1} = 2$ happens if we lose \$1 when our capital is a multiple of 3. According to the rules, we flip Coin 1 when the capital is a multiple of 3. Losing \$1 with Coin 1 happens with probability $1 - (1/10 - \epsilon) = 9/10 + \epsilon$.

To show that Game B is a losing game, we need to find the steady-state vector π for the Markov chain X_n. The calculations will be simpler if we assigned a number to ϵ. To this end, let's pick $\epsilon = 0.01$. The transition matrix P simplifies as

$$\begin{bmatrix} 0 & 0.09 & 0.91 \\ 0.26 & 0 & 0.74 \\ 0.74 & 0.26 & 0 \end{bmatrix}.$$

To find π, we need to solve the equations $\pi P = \pi$ and $\pi(0) + \pi(1) + \pi(2) = 1$. This gives us the system of equations

$$0.26\pi(1) + 0.74\pi(2) = \pi(0)$$
$$0.09\pi(0) + 0.26\pi(2) = \pi(1)$$
$$0.91\pi(0) + 0.74\pi(1) = \pi(2)$$
$$\pi(0) + \pi(1) + \pi(2) = 1.$$

One of the top three equations is redundant. We rewrite the first, second, and fourth equations as

$$-\pi(0) + 0.26\pi(1) + 0.74\pi(2) = 0$$
$$0.09\pi(0) - \pi(1) + 0.26\pi(2) = 0$$
$$\pi(0) + \pi(1) + \pi(2) = 1.$$

We solve these equations in Julia and find $\pi = [0.38, 0.16, 0.46]$. This means in the long run the Markov chain X_n will be in state zero 38% of the time, in state one 16% of the time, and in state two 46% of the time. Then, we can compute $P(\text{win \$1})$ by conditioning on the state of the Markov chain in the long run:

$$P(\text{win \$1}) = \pi(0)p_{01} + \pi(1)p_{12} + \pi(2)p_{20}$$
$$= 0.38 \times 0.09 + 0.16 \times 0.74 + 0.46 \times 0.74$$
$$= 0.49 < 1/2.$$

Therefore Game B is a losing game.

4.3.2.3 Game C

Here comes the mystery Game C! The description of the game is simple. Use a fair coin, and randomly switch between Game A and Game B. In other words, flip a fair coin at each time n, and if it lands "heads" play Game A, otherwise play Game B. We want to compute $P(\text{win } \$1)$ for Game C, which will be done following the same approach we used for Game B.

As in Game B, let X_n denote the remainder when our capital is divided by three at the nth time we play the game. The transition matrix for X_n is

$$P = \begin{bmatrix} p_{00} & p_{01} & p_{02} \\ p_{10} & p_{11} & p_{12} \\ p_{20} & p_{21} & p_{22} \end{bmatrix}.$$

Note that $p_{00} = p_{11} = p_{22} = 0$ since we either win or lose each time we play. Let's compute p_{01} and p_{12}, and leave the rest to the reader.

We have

$$\begin{aligned} p_{01} &= P(X_{n+1} = 1 | X_n = 0) \\ &= P(\text{win } \$1) \\ &= P(\text{win } \$1 | \text{Game A is played}) P(\text{Game A is played}) \\ &\quad + P(\text{win } \$1 | \text{Game B is played}) P(\text{Game B is played}) \\ &= \left(\frac{1}{2} - \epsilon \right) \frac{1}{2} + \left(\frac{1}{10} - \epsilon \right) \frac{1}{2}, \end{aligned}$$

which is 0.29 if $\epsilon = 0.01$.

Similarly,

$$p_{12} = P(X_{n+1} = 2 | X_n = 1) = P(\text{win } \$1) = \left(\frac{1}{2} - \epsilon \right) \frac{1}{2} + \left(\frac{3}{4} - \epsilon \right) \frac{1}{2},$$

which is 0.615 if $\epsilon = 0.01$.

After we compute the rest of the transition matrix we obtain, for $\epsilon = 0.01$,

$$P = \begin{bmatrix} 0 & 0.29 & 0.71 \\ 0.385 & 0 & 0.615 \\ 0.615 & p_{21} & 0 \end{bmatrix}.$$

We then find the steady-state vector π, following similar calculations as in Game B:

$$\pi = [0.345, 0.254, 0.401].$$

Next we compute the probability of winning and find

$$P(\text{win } \$1) = \pi(0)p_{01} + \pi(1)p_{12} + \pi(2)p_{20}$$
$$= 0.345 \times 0.29 + 0.254 \times 0.615 + 0.401 \times 0.615$$
$$= 0.503 > 1/2.$$

Rather remarkably, Game C, obtained from two losing games, is itself a winning game! This is Parrando's paradox!

4.4 Project 12: **Market share of shoe brands**

There is a recent development in Tallahassee—a strip mall with four shoe stores! The stores are Adidas, Nike, Converse, and Shoe Station. Shoe Station will carry brands other than Adidas, Nike, and Converse. What is interesting about the strip mall is that the shoe stores are planning to build a basketball court in an attempt to support youth sports as well as promote their products. The initial cost of the basketball court is $50,000, and then there is the yearly maintenance of $10,000. The four stores agree to share the initial cost equally, but not the yearly maintenance cost. The manager of Shoe Station wants the yearly cost to be divided in proportion to the market share of the stores.

We will model this problem using Markov chains, and find the percentage of customers shopping at each store in the long run, which is the information needed to divide the yearly cost of the basketball court among the stores in a proportional way. The states of the Markov chain are A (Adidas), N (Nike), C (Converse), and S (Shoe Station).

1. Think about the brand of the pair of shoes you own now, and what brand you want to purchase next. Collect this information from your friends.[2] We will use this data as a proxy for the real customers of the stores. Obtain transition probabilities using this data. For example, if 5 students own Nike shoes, and 2 of them want to buy Nike again as their next pair of shoes, then $2/5 = 40\%$ is the transition probability from N to N, that is, $p_{NN} = 2/5$. (If unable to generate data in this way, use the artificial data given in Appendix B.)

---

[2] We survey YSP students in the class for this project.

2. Find the steady-state vector π, if it exists. Find how much each store should pay for the yearly maintenance.
3. Investigate the sensitivity of the steady-state vector to the transition probabilities. For example, if the estimate for p_{AA} you had in part (1) changed a little bit, how much would the steady-state vector change? Use Julia to consider several scenarios and analyze the sensitivity empirically. (You can simplify this question by concentrating on the sensitivity of just one component of the steady-state vector.)

Chapter 5
Brownian Motion

In the previous chapter, we learned about Markov chains: a sequence of random variables $X_0, X_1, \ldots$ with a specific property (usually called the Markovian property) described in Definition 4.1. We often think about the subscript of the random variables X as time: X_0 is the current time, and time increments in steps of one unit which could be one second, hour, year, etc. This is a "discrete" representation of time, and for that reason the Markov chains we studied in the previous chapter are called **discrete Markov chains**.

Time can also be modeled using a continuous variable t, like in $X(t)$ where $t \geq 0$ is a real number. Thinking about time in this way has certain advantages. In the iPhone zombie Example 4.1, we measured the location of the student by the number of steps he took; this was a discrete way of measuring time. If we switch to continuous time, we can talk about the location of the student at any time t.

A collection of random variables where the index is a continuous variable, written as $\{X(t), t \in T\}$ where T is usually a finite or an infinite interval of the real line, is called a **stochastic process**. In this chapter we will discuss two stochastic processes, the **Brownian motion** and the **geometric Brownian motion**.

5.1 Brownian motion

The name Brownian motion comes from Robert Brown, a botanist who observed in 1827, under a microscope, that grains of pollens suspended in water displayed a continuous wiggly motion, similar to the wiggly motion plotted in Figure 5.1.

Brown wrote [6]:

> Having found motion in the particles of the pollen of all the living plants which I had examined, I was led next to inquire whether this property continued after the death of the plant, and for what length of time it was retained. In plants, either dried or immersed in spirit for a few days only, the particles of pollen of both kinds were found in motion equally evident

© The Editor(s) (if applicable) and The Author(s), under exclusive license to Springer Nature Switzerland AG 2020
G. Ökten, *Probability and Simulation*, Springer Undergraduate Texts in Mathematics and Technology, https://doi.org/10.1007/978-3-030-56070-6_5

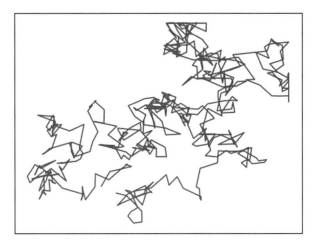

Fig. 5.1: Brownian motion

with that observed in the living plant; specimens of several plants, some of which had been dried and preserved in an herbarium for upwards of twenty years, and others not less than a century, still exhibited the molecules or smaller spherical particles in considerable numbers, and in evident motion, along with a few of the larger particles, whose motions were much less manifest, and in some cases not observable.

A mathematical analysis of the motion observed by Robert Brown was made by Albert Einstein in 1905. One of the first applications of Brownian motion was to model stock prices, which was established by Louis Bachelier in his doctoral dissertation in 1900.

Here is the modern mathematical definition of Brownian motion.

Definition 5.1 (Brownian motion) The stochastic process $\{W(t); 0 \leq t \leq T\}$ is a Brownian motion with parameters μ and σ on $[0, T]$ if

1. $W(0) = 0$.
2. The process has independent increments: for any $t_1 < t_2 < \ldots < t_n$, $W(t_2) - W(t_1), W(t_3) - W(t_2), \ldots, W(t_n) - W(t_{n-1})$ are independent random variables.
3. For every $0 \leq s < t \leq T$, $W(t) - W(s) \sim N(\mu(t - s), \sigma^2(t - s))$.

The process is denoted by $BM(\mu, \sigma^2)$. The parameter μ is called the drift, and σ^2 the variance of the Brownian motion. The special case $\mu = 0, \sigma^2 = 1$ is called the **standard Brownian motion**.

Let's dissect this definition. The first condition simply says our infinite collection of random variables start at the origin: $W(0)$ is fixed at the constant 0. The second and third conditions are about the increments, $W(t) - W(s)$ for $s < t$. Keep in mind that the increments are themselves random variables. In a way, we are defining the stochastic process in terms of its increments. The third condition says an increment is a normal random variable with mean μ times the time increment, and with variance equal to σ^2 times the time increment. The second condition says the increments over nonoverlapping time intervals are independent random variables.

What does Brownian motion look like? We can get a visualization by simulating the process. By that we mean the following: specify a fixed set of points in time, $0 < t_1 < \ldots < t_n$, and simulate the random variables $W(t_1), \ldots, W(t_n)$. We will set $t_0 = 0$, and thus $W(t_0) = 0$. To simulate the aforementioned values, we will use the third condition of Definition 5.1 which says the random variables

$$W(t_1) - W(t_0), W(t_2) - W(t_1), \ldots, W(t_n) - W(t_{n-1})$$

are independent, and have distribution $N(\mu(t_i - t_{i-1}), \sigma^2(t_i - t_{i-1}))$ as $i = 1, \ldots, n$.

Observe that generating a value from $N(\mu(t_i - t_{i-1}), \sigma^2(t_i - t_{i-1}))$ can be done by generating a random number z from the standard normal distribution $N(0, 1)$ and then computing

$$\mu(t_i - t_{i-1}) + \sigma \sqrt{t_i - t_{i-1}}\, z.$$

Therefore, to simulate the increments, we generate n independent random numbers $z_1, \ldots, z_n$ from the standard normal distribution, and set

$$W(t_1) - W(t_0) = W(t_1) = \mu(t_1 - t_0) + \sigma \sqrt{t_1 - t_0}\, z_1 \qquad (5.1)$$
$$W(t_2) - W(t_1) = \mu(t_2 - t_1) + \sigma \sqrt{t_2 - t_1}\, z_2$$
$$\cdots$$
$$W(t_n) - W(t_{n-1}) = \mu(t_n - t_{n-1}) + \sigma \sqrt{t_n - t_{n-1}}\, z_n$$

Solving these equations recursively for $W(t_1), W(t_2), \ldots, W(t_n)$ gives us the Brownian motion path.

We will use Julia to simulate and plot Brownian motion paths. For simplicity, we will consider equal spacing of time: $t_0 = 0, t_1 = 1/n, \ldots, t_i = i/n, \ldots, t_n = 1$ where n is a parameter we will pick. With this, Eq. (5.1) simplifies as

$$W(t_1) = \mu/n + \sigma \sqrt{1/n}\, z_1$$
$$W(t_2) = W(t_1) + \mu/n + \sigma \sqrt{1/n}\, z_2$$
$$\cdots$$

In the Julia code below, we will plot Brownian motion paths with increasing values of n to see how the parameter n affects the behavior of the process.

5.1.1 Simulating Brownian motion

```
In [1]: using PyPlot

In [2]: function BrownianMotion(mu,sigma,n)
            values=zeros(n+1)
            values[1]=0
            for i in 2:n+1
```

```
            w=values[i-1]+mu/n+sigma*((1/n)^0.5)*randn()
            values[i]=w
        end
        xvalues=0:1/n:1
        plot(xvalues,values)
    end
```

Out[2]: BrownianMotion (generic function with 1 method)

In [3]: BrownianMotion(0,0.2,10);

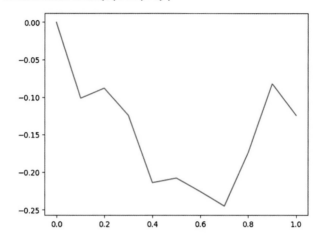

In [4]: BrownianMotion(0,0.2,100);

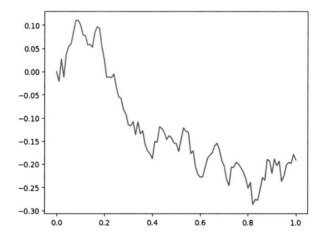

In [5]: BrownianMotion(0,0.2,1000);

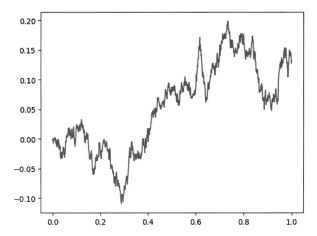

Notice that as n gets larger and the time interval $(0, 1)$ is divided into smaller and smaller subintervals, the Brownian motion path becomes more wiggly.

5.2 Project 13: Modeling insect movement

An important research problem in ecology is developing mathematical models for the movement behavior of animals. A quantitative model predicting the location of an animal can be useful in conservation biology and pest control. In this project, we will discuss a simple model based on two-dimensional Brownian motion. For a more realistic generalization of this model, see Kareiva and Shigesada [15].

Imagine an insect in a two-dimensional grid, with initial position at the origin. The insect then starts moving in a way which would likely look erratic to our eyes. Let $(X(t), Y(t))$ be the location, that is, the x- and y-coordinates of the insect at time t, with $X(0) = Y(0) = 0$. Our model assumes $X(t)$ and $Y(t)$ are two independent standard Brownian motions.

Our first task is to simulate the model and obtain trajectories for the insect. To do that we need to discretize time: let $t_0 = 0, t_1, t_2, \ldots$ be the discrete times of insect movement, and for simplicity let's set the time difference between two consecutive moves to one unit of time. We can then switch to a simpler notation and set $X(t_0) = X(0), X(t_1) = X(1)$ and so forth, and similarly set $Y(t_i) = Y(i)$.

1. By updating Eq. (5.1), show that the movement of the insect can be simulated using the following equations, where $x(i), y(i)$ correspond to particular values the random variables $X(i), Y(i)$ take:

$$x(i) = x(i-1) + z_1$$
$$y(i) = y(i-1) + z_2 \qquad\qquad (5.2)$$

where z_1, z_2 are independent random numbers from the standard normal distribution. Consider the grid $[-10, 10] \times [-10, 10]$, with the insect located at the origin. Write a Julia code that plots the movement of the insect. If the insect's x- or y-coordinates go beyond the borders $x, y = \pm 10$, assume it stays on the border. For example, if the x-coordinate exceeds 10 at the fourth move, that is $x(4) > 10$, then set $x(4) = 10$. Your code should take the number of total moves n as an input, and plot $(x(0), y(0)), (x(1), y(1)), \ldots, (x(n), y(n))$ as the output. Plot two trajectories using $n = 200$ and $n = 400$ moves.

2. Let $L(i)$ be the random variable that gives the length of the ith move of the insect. We will denote a particular value from this random variable as $l(i)$. Update your Julia code to estimate the average length of a move, $\frac{1}{n} \sum_{i=1}^{n} l(i)$, and the average length squared, $\frac{1}{n} \sum_{i=1}^{n} l^2(i)$. Do these averages seem to converge to a value as n gets larger?

3. The average length squared can be computed analytically. Use the properties of Brownian motion to prove $E[L(i)^2] = 2$ for any i.

4. Now assume there is food for the insect at the top-right corner of the grid, at $(10, 10)$, and the insect can smell it once in enters the rectangle $[5, 10] \times [5, 10]$. Once in this rectangle, the insect will only move toward the top-right corner. How can you modify Eq. (5.2), in particular the random numbers in the equation, so that $x(i)$ and $y(i)$ will only increase once the insect is in the region? Plot new

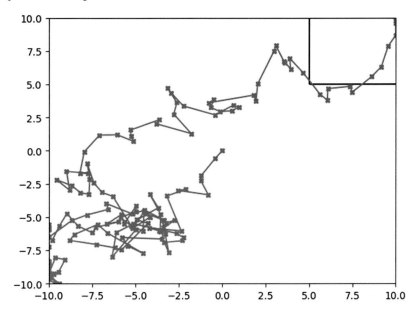

trajectories for the insect until it finds the food. Below is an example of such a trajectory.

5. Update your code so that it will simulate N trajectories until the insect finds the food, compute the number of moves until the insect finds the food for each trajectory, and then return the average number of moves to find the food over N trajectories. Simulate $N = 1000$ and $N = 10000$ trajectories to find the average number of moves.

5.3 Geometric Brownian motion

Figure 5.2 plots the daily opening stock price (the price the stock trades for when an exchange opens for the day) for Apple Inc. from 10/1/2018 to 10/1/2019. The erratic behavior of the stock price resembles the ragged movement of Brownian motion. Could Brownian motion be a good model for the behavior of stock prices?

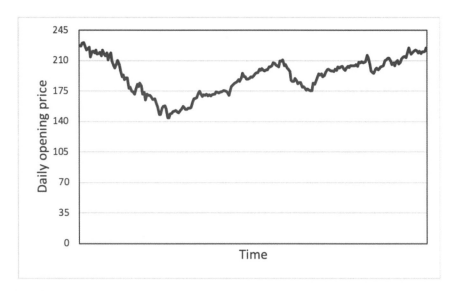

Fig. 5.2: Daily opening stock price of Apple Inc. between 10/1/2018 and 10/1/2019

In his doctoral dissertation published in 1900[1], Louis Bachelier indeed used Brownian motion to model stock prices. Let $S(t)$ denote the price of a stock where t is time, and let t_0 be the current time. Bachelier assumed $\{S(t) - S(t_0), t \geq 0\}$ is a Brownian motion with parameters μ and σ.

From the definition of Brownian motion, Bachelier's model implies the stock price increments $S(t_3) - S(t_2)$ and $S(t_2) - S(t_1)$ are independent, for any $t_1 < t_2 < t_3$, and the distribution of the stock price difference $S(t_2) - S(t_1)$ is $N(\mu(t_2 - t_1), \sigma^2(t_2 - t_1))$ for any time points $t_1 < t_2$.

Bachelier's model has some flaws. One is the fact that $S(t)$ can be negative under this model. To describe the other flaw, consider these two events:

- Stock price increases from \$10 to \$20 from time t_1 to t_2 (100% increase),
- Stock price increases from \$100 to \$110 from time t_1 to t_2 (10% increase).

In each event the difference $S(t_2) - S(t_1)$ has the same distribution: $N(\mu(t_2 - t_1), \sigma^2(t_2 - t_1))$. Therefore the probability that this difference is \$10, which is the case

---

[1] Bachelier, L., 1900. Théorie de la spéculation. In Annales scientifiques de l'École normale supérieure (Vol. 17, pp. 21–86).

in each event, must be the same.[2] But this is very unrealistic: a 10% increase in the stock price will certainly not be as likely as a 100% increase!

Paul Samuelson, a Nobel laureate in economics, modified Bachelier's model to address these flaws in a paper he wrote in 1965 [22]. Instead of assuming

$$\{S(t) - S(t_0), t \geq 0\}$$

is a Brownian motion, he assumed

$$\left\{\log \frac{S(t)}{S(t_0)}, t \geq 0\right\}$$

is a Brownian motion with some parameters μ, σ. This model implies the logarithm of the stock price ratios are normally distributed, that is

$$\log \frac{S(t)}{S(t_0)} \sim N(\mu(t - t_0), \sigma^2(t - t_0)).$$

The choice of time parameters t_0, t is arbitrary; we can write the above statement as

$$\log \frac{S(t_2)}{S(t_1)} \sim N(\mu(t_2 - t_1), \sigma^2(t_2 - t_1)) \tag{5.3}$$

for any $t_1 < t_2$. This stochastic process is called a **geometric Brownian motion** with parameters μ and σ. The parameter μ is called the drift, and σ is called the volatility of the stock.

Definition 5.2 (Geometric Brownian motion) A stochastic process $\{S(t), 0 \leq t \leq T\}$ is a geometric Brownian motion with parameters μ, σ, if $\{\log \frac{S(t)}{S(0)}, t \leq t \leq T\}$ is a Brownian motion with parameters μ, σ.

To answer questions about geometric Brownian motion, we need to understand more about random variables that appear in Eq. (5.3): these are random variables whose logarithm is a normal random variable. Such random variables are called **lognormal random variables**.

Definition 5.3 (Lognormal random variable) We say X is a lognormal random variable with parameters m, s^2 if $\log X \sim N(m, s^2)$.

The expected value and variance of the lognormal random variable with parameters m, s^2 is

$$E[X] = e^{m+s^2/2}, \ Var(X) = e^{2m+s^2}\left(e^{s^2} - 1\right), \tag{5.4}$$

a proof of which can be found in most probability books. Now we are ready to answer some interesting questions.

---

[2] Actually this probability is zero since we are dealing with continuous random variables, so a better way to pose the question would be to specify an interval for the stock price in each event, instead of the fixed price $10.

Example 5.1 (Growth of stock price) Assume we have a stock, whose price is modeled using Samuelson's model, with $\mu = 0.1$ and $\sigma = 0.2$. Assume today's stock price is \$20. What is the expected value of the stock price after 6 months, that is, when $t = 0.5$? (t is measured in years.) What is its variance?

Solution 5.1 Let $S(t)$ denote the stock price at time t with $S(0) = 20$. From Eq. (5.3) we have

$$\log \underbrace{\frac{S(t)}{S(0)}}_{X} \sim N(\mu t, \sigma^2 t) = N(0.1t, 0.04t).$$

Notice that $X = \frac{S(t)}{S(0)}$ is a lognormal random variable with parameters $0.1t$ and $0.04t$. To find its mean and variance, we need to substitute $m = 0.1t$, $s^2 = 0.04t$ in Eq. (5.4):

$$E\left[\frac{S(t)}{S(0)}\right] = e^{0.12t}, \quad Var\left(\frac{S(t)}{S(0)}\right) = e^{0.24t}(e^{0.04t} - 1).$$

Finally, substitute $t = 0.5$ to get

$$E[S(0.5)] = 20e^{0.06} = 21.2, \quad Var(S(0.5)) = (20)^2 e^{0.12}(e^{0.02} - 1) = 9.1.$$

5.3.1 Simulating stock prices

We will simulate Samuelson's model for the stock price: the geometric Brownian motion described by Eq. (5.3). Let's label today as $t_0 = 0$. We want to simulate the price of the stock at times $t_1, t_2, \ldots, t_m = T$, where $t_m = T$ is some time in the future. We can simplify the choice of the time points by assuming a uniform time increment, say h, so that starting with today, the time points are $t_0 = 0, t_1 = h, t_2 = 2h, \ldots$, $t_m = mh = T$. Note that there are m time intervals and each time interval has length h (see the diagram below).

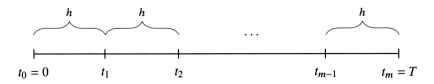

To obtain the value of $S(t_1) = S(h)$, we use Eq. (5.3):

$$\log \frac{S(h)}{S(0)} \sim N(\mu h, \sigma^2 h).$$

We can rewrite this as

$$\log S(h) = \log S(0) + \mu h + \sigma Z_1 \sqrt{h}$$

where Z_1 is a standard normal random variable.

Similarly, we have

$$\log \frac{S(2h)}{S(h)} \sim N(\mu h, \sigma^2 h)$$

and thus

$$\log S(2h) = \log S(h) + \mu h + \sigma Z_2 \sqrt{h}$$

where Z_2 is a standard normal random variable. Note that Z_1, Z_2 are independent random variables. The pattern should be clear: to simulate a stock price path $S(h), S(2h), \ldots, S(mh)$, generate m random numbers from the standard normal distribution, $z_1, \ldots, z_m$, and set

$$\log S(h) = \log S(0) + \mu h + \sigma z_1 \sqrt{h}$$
$$\log S(2h) = \log S(h) + \mu h + \sigma z_2 \sqrt{h}$$
$$\ldots \tag{5.5}$$
$$\log S(mh) = \log S((m-1)h) + \mu h + \sigma z_m \sqrt{h}$$

These equations simulate the values of the logarithm of the stock price. If we need the stock price itself, we simply compute $e^{\log(S)}$.

In Example 5.1 we computed the mean and variance of the stock price. Let's use Monte Carlo simulation to estimate the mean of the stock price, and check how close the estimate is to the correct value. First we write a Julia code that implements Eqs. (5.5). The code takes the number of stock price paths to generate, N, as one of the inputs. It computes the stock price $S(t_m = T)$ for each path, and then computes the average of these prices (**finalprice/N** in the code), which is an approximation for $E[S(t_m)]$.

```
In [1]: function stockprices(mu,sigma,szero,m,T,N)
            finalprice=0
            h=T/m
            for j in 1:N
                logprices=zeros(m+1)
                logprices[1]=log(szero)
                for i in 2:m+1
                    logprices[i]=logprices[i-1]+mu*h
                            +sigma*(h^0.5)*randn()
                end
                finalprice=finalprice+exp(logprices[m+1])
            end
            finalprice/N
        end

Out[1]: stockprices (generic function with 1 method)
```

Next we enter the parameters used in Example 5.1 as the inputs to the function **stockprices**. We use $m = 10$, which means the time interval $(0, T = 0.5)$ is divided into 10 subintervals, and we generate $N = 10000$ stock price paths. The Monte Carlo estimate for the mean of the stock price at the expiry, $E[S(0.5)]$, is

```
In [2]: stockprices(0.1,0.2,20,10,0.5,10000)
Out[2]: 21.227114681683545
```

which is pretty close to the theoretical value 21.2 calculated in Example 5.1.

Let's plot one of the stock price paths used in the simulation. We need to modify the Julia code above so that we compute the actual stock price instead of its logarithm at each step. Raising both sides to power e in Eqs. (5.5), we obtain

$$S(h) = S(0) \exp(\mu h + \sigma \sqrt{h} z_1)$$
$$S(2h) = S(h) \exp(\mu h + \sigma \sqrt{h} z_2)$$
$$\ldots \tag{5.6}$$
$$S(mh) = S((m-1)h) \exp(\mu h + \sigma \sqrt{h} z_m)$$

We implement these equations in Julia.

```
In [3]: using PyPlot
In [4]: function stockpricepath(mu,sigma,szero,m,T)
            h=T/m
            prices=zeros(m+1)
            prices[1]=szero
            for i in 2:m+1
                prices[i]=prices[i-1]*exp(mu*h+sigma*(h^0.5)
                    *randn())
            end
            xvalues=0:1/(2*m):0.5
            plot(xvalues,prices)
            xlabel("Time")
            ylabel("Stock price")
        end

Out[4]: stockpricepath (generic function with 1 method)
```

Here is the plot of a stock price path.

```
In [5]: stockpricepath(0.1,0.2,20,100,0.5);
```

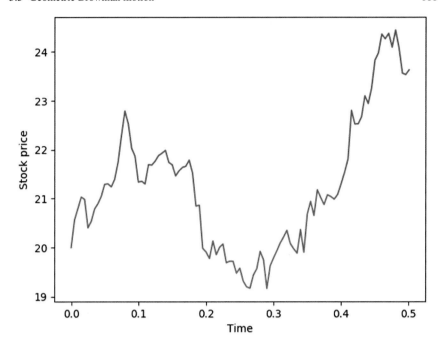

5.4 Project 14: Option pricing

A European call option is an example of a *financial derivative*. You can buy and sell a European call option, just like a stock. A call option is tied to an *underlying*, which is usually a stock. For example, consider the Amazon.com stock which is traded for $1,733 today. We can buy a European call option on this stock, with a strike price of $1,715 and an expiry of four days, for the price of $30. At the end of the expiry (after four days), the call option gives us the right to purchase the Amazon.com stock for $1,715 (the strike price), no matter how much it is trading for on that day. For example, if the stock trades for $1,730 at the expiry, then we can make a profit of $15 by buying the stock at $1,715 and selling it for $1,730. If the stock trades for $1,700 at the expiry, then we *would not* strike the call option and purchase the stock for $1,715, since it is traded for a cheaper price. In that case, we would be losing our initial investment of $30 that we used to purchase the call option. A famous question in financial mathematics was how to find the fair price of a European call option. The answer was given by the celebrated Black–Scholes–Merton formula, for which Robert Merton and Myron Scholes received the Nobel prize in Economics in 1997.

Let's introduce some notation. Let C be the price of the European call option, K the strike price, and $S(t)$ the underlying stock price at t. We denote the expiry by T. Let r denote the risk-free interest rate; this is like the interest rate a savings account pays. The payoff of the call option is $\max(S(T) - K, 0)$: that is, the option payoff is $S(T) - K$ if $S(T) > K$, and otherwise it is zero. We assume the stock price follows the geometric Brownian motion model with drift μ and volatility σ.

An important result from financial mathematics states that the fair price of the call option is

$$C = e^{-rT} E[\max(S(T) - K, 0)], \tag{5.7}$$

where $S(T)$ follows the geometric Brownian motion with drift $r - \sigma^2/2$ and volatility σ:

$$\log \frac{S(T)}{S(0)} \sim N((r - \sigma^2/2)T, \sigma^2 T). \tag{5.8}$$

Here are some interesting observations about this result:

- Eq. (5.7) states the price of the call option is the expected value of its payoff, discounted to today's dollars: note that $1 today is e^{rT} dollars at time T since the money grows at risk-free interest rate r via continuous compounding. Equivalently, $1 at time T is worth e^{-rT} dollars today.
- Even though we originally assumed $S(T)$ followed geometric Brownian motion with drift μ and volatility σ, in Eq. (5.8) the drift term was replaced by $r - \sigma^2/2$. The volatility term remained the same. This fascinating phenomenon is known as **risk-neutral pricing**, a topic covered in textbooks on financial mathematics.

In this project we will pick a call option that is traded in the markets, compute its fair price using the geometric Brownian motion model, and compare the price with the price it is trading for in the market.

1. Look up resources on the web such as Yahoo! Finance to find information about a call option on the General Electric (GE) stock[3]. Find out the following values: today's date t_0, expiry T, strike price K, the last price the option was traded for, $\tilde{C}$, and the last price the stock was traded for, S_0. Then find the interest rate for a one-year certificate of deposit: we will use this value as the risk-free interest rate r.

2. We will estimate the volatility σ of the GE stock from historical data. Download historical daily stock price data for GE[4]. There are 252 trading days in a year, so one day amounts to 1/252 years. The geometric Brownian motion model Eq. (5.3) takes the following form when $t_1, t_2, \ldots$, represent consecutive trading days:

$$\log \frac{S(t_i)}{S(t_{i-1})} \sim N(\mu/252, \sigma^2/252) \tag{5.9}$$

for $i = 1, 2, \ldots$. Let's label the downloaded stock price data as $S_1, S_2, \ldots, S_m$ for some m. Use this data to compute

$$\log\left(\frac{S_2}{S_1}\right), \log\left(\frac{S_3}{S_2}\right), \ldots, \log\left(\frac{S_m}{S_{m-1}}\right),$$

and then compute the sample variance $\hat{\sigma}^2$ of these ratios. We have

$$\hat{\sigma}^2 \approx \sigma^2/252$$

from Eq. (5.9). Therefore we estimate the volatility by $\sigma = \sqrt{252}\hat{\sigma}$.

3. Use Monte Carlo simulation to estimate the price of the call option whose parameters you found in parts (1) and (2), using Eqs. (5.7) and (5.8). To do this, you will simulate N stock prices $S^{(1)}(T), \ldots, S^{(N)}(T)$ using the model described by Eq. (5.8), and estimate C using the sample average

$$\frac{e^{-rT}}{N} \sum_{i=1}^{N} \max(S^{(i)}(T) - K, 0).$$

Use $N = 1000$ and $N = 10000$. Compare your estimates for C with the price the option is traded for, $\tilde{C}$. Answer this question using call options with different strike prices, fixing all other parameters. Are C and $\tilde{C}$ closer to each other for some strike prices than others?

---

[3] See, for example, the "options" link at https://finance.yahoo.com/quote/GE/.
[4] This can be obtained from the "historical data" link at https://finance.yahoo.com/quote/GE/.

Appendix A
Benford's law

We will use a package called JuliaDB to import data. After running Pkg.add("JuliaDB"), we can load it:

In [1]: `using PyPlot;`

In [2]: `using JuliaDB;`

The data we will study is the US population by county. It was obtained from US Census Bureau (factfinder.census.gov). The data was saved in the same directory where this Julia notebook is as a CSV file. The name of the file is "population_county.csv". The function loadtable will import the data to a table. If there are any errors at this stage, open the CSV file and look for entries that are not numbers.

In [3]: `data=loadtable("population_county.csv")`

Out[3]: Table with 3142 rows, 14 columns:
Columns:

1	GEO.id	String
2	GEO.id2	Int64
3	GEO.display-label	String
4	rescen42010	Int64
5	resbase42010	Int64
6	respop72010	Int64
7	respop72011	Int64
8	respop72012	Int64
9	respop72013	Int64
10	respop72014	Int64
11	respop72015	Int64
12	respop72016	Int64
13	respop72017	Int64
14	respop72018	Int64

G. Ökten, *Probability and Simulation*, Springer Undergraduate Texts in Mathematics and Technology, https://doi.org/10.1007/978-3-030-56070-6

The function select will pick any column of the data and store it as an array. The second argument is the title of the column.

```
In [4]: pop=select(data, :rescen42010);
```

To extract the first number in the list, we type

```
In [5]: pop[1]
```

```
Out[5]: 54571
```

We want the first digit of this number. Let's see what the function digits does.

```
In [6]: digits(pop[1])
```

```
Out[6]: 5-element Array{Int64,1}:
         1
         7
         5
         4
         5
```

The first digit, that is the most significant digits, appears at the end of the output of **digits**. We can reverse the ordering by

```
In [7]: reverse(digits(pop[1]))
```

```
Out[7]: 5-element Array{Int64,1}:
         5
         4
         5
         7
         1
```

The function **first** picks the first number of an array.

```
In [8]: first(reverse(digits(pop[1])))
```

```
Out[8]: 5
```

The following for loop goes through the array pop and picks the first digit of each number.

```
In [9]: first_digs=[first(reverse(digits(n))) for n in pop];
```

A histogram of the first digits comes next:

```
In [10]: hist(first_digs,9)
         xlabel("Digits")
         ylabel("Frequency");
```

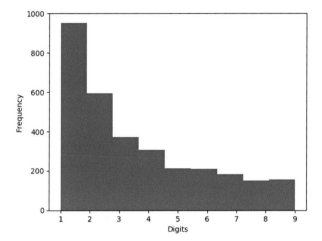

Clearly the earlier digits occur more often than the higher ones. This is what Benford observed in 1938, and Newcomb in 1881.

What about the distribution of other digits? Let's see what the distribution of third digits looks like. In the data, however, there are a few numbers with only two digits. We remove those numbers from the original data set, and call the new data set "population_county_threedigs.csv". (The data is available on the Springer web page for the book.) Next we load this data, and plot a histogram for the third digits. Note that the third digit can be any integer between 0 and 9, so in the histogram we will specify 10 bins.

```
In [11]: data=loadtable("population_county_threedigs.csv")
         pop=select(data, :rescen42010)
         first_digs=[reverse(digits(n))[1] for n in pop];
         sec_digs=[reverse(digits(n))[2] for n in pop];
         third_digs=[reverse(digits(n))[3] for n in pop];
         xlabel("First digits")
         ylabel("Relative frequency")
         hist(first_digs,density=true,9);
```

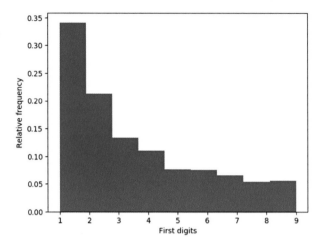

```
In [12]: hist(sec_digs,10,density=true);
         xlabel("Second digits")
         ylabel("Relative frequency");
```

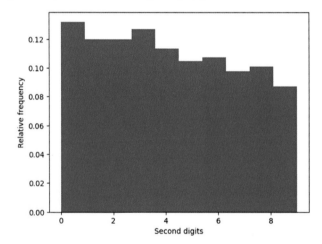

```
In [13]: hist(third_digs,10,density=true);
         xlabel("Third digits")
         ylabel("Relative frequency");
```

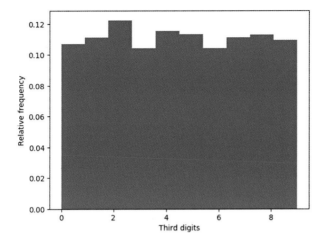

This histogram looks pretty uniform. Let's compare this with the histogram of 3140 random numbers (this is how many numbers we had in the array third_digs) from the discrete uniform distribution on the integers 0,1,...,9. The function **rand** can be used to generate random integers in a given range. Here is how to get 20 such random integers:

```
In [14]: rand(0:9,5);

Out[14]: 5-element Array{Int64,1}:
         4
         9
         6
         6
         3
```

Here is a histogram for 3140 random numbers from the uniform distribution on 0,1,...,9:

```
In [15]: hist(rand(0:9,3140),10,density=true);
         xlabel("Uniform random integers 0 through 9")
         ylabel("Relative frequency");
```

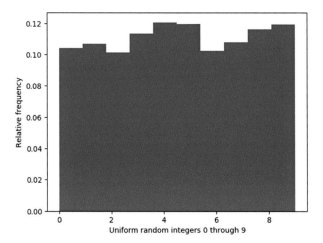

This is visually indistinguishable from the histogram of the third digits. Although the distribution of the first digits is very different than uniform, as we look at the later digits, the distribution quickly becomes uniform. This observation was made by Newcomb as well.

Let's go back to the distribution of the first digits. Benford's law gives the probability of the first digit as

$$\text{Prob(first digit } = d) = \log_{10}(1 + 1/d).$$

We write a Julia code for this function.

```
In [16]: p(d)=log(10,1+1/d)
```

```
Out[16]: p (generic function with 1 method)
```

The probability that the first digit is a 1 is given by

```
In [17]: p(1)
```

```
Out[17]: 0.30102999566398114
```

Let's plot the Benford probability mass function together with the histogram to see how good the fit is. There is one problem though. The y-axis of the histogram is the frequency of the digits, which goes as high as 1000. We need to normalize the heights so that they are between 0 and 1. This is done using the option "density=true" in the **hist** function below.

```
In [18]: xaxis=1:.1:9
         yvals=map(t->p(t),xaxis)
         plot(xaxis,yvals,label="Benford density")
         xlabel("Digits")
         ylabel("Relative frequency")
         hist(first_digs,9,density=true,label="Population data")
         legend(loc="upper center");
```

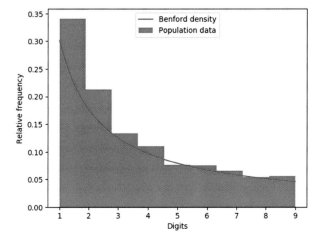

Appendix B
Data for Project 12: Market share of shoe brands

This is an artificial data set for 100 students, and their choices for their next pair of shoes. The states A, N, C, S, correspond to Adidas, Nike, Converse, and Shoe Station. The number 7 in the first row means 7 students who own Adidas shoes will buy Nike for their next pair of shoes.

Current \ Next	A	N	C	S
A	5	7	2	1
N	5	10	4	2
C	8	2	3	3
S	15	15	8	10

To obtain the transition probability of switching from Adidas to Nike, divide 7 by the row sum: $P_{AN} = 7/15$.

Solutions

Project 1: Verifying identities using Julia

The probability that **verifyid** returns "the polynomials are equivalent" when in fact they are not equivalent is at most $(\frac{1}{100})^{10}$.

```
In [1]: function verifyid(F::Function,G::Function,d)
            flag=true
            n=1
            while (flag==true) && (n <=10)
                r=rand(1:100*d)
                flag=(F(r)==G(r))
                n=n+1
            end
            println("n is $n")
            if flag==true
                print("The two polynomials are equivalent")
            else
                print("The two polynomials are not equivalent")
            end
        end

Out[1]: verifyid (generic function with 1 method)

In [2]: F(x)=(x-5)*(x-10)*(x+3)*(x-2)*(x+25)

Out[2]: F (generic function with 1 method)

In [3]: G(x)=x^5+11*x^4-321*x^3+865*x^2+3200*x-7500

Out[3]: G (generic function with 1 method)
```

© The Editor(s) (if applicable) and The Author(s), under exclusive license to Springer Nature Switzerland AG 2020
G. Ökten, *Probability and Simulation*, Springer Undergraduate Texts in Mathematics and Technology, https://doi.org/10.1007/978-3-030-56070-6

The above expressions are equivalent from WolframAlpha.

```
In [4]: verifyid(F,G,5)
```

```
n is 11
The two polynomials are equivalent
```

Now let's pick a wrong polynomial:

```
In [5]: H(x)=x^5+11*x^4-321*x^3+865*x^2+3200*x-7501
```

```
Out[5]: H (generic function with 1 method)
```

```
In [6]: verifyid(F,H,5)
```

```
n is 2
The two polynomials are not equivalent
```

Project 2: Analysis of Freivalds' algorithm

1. • Generate $r^{(1)}, r^{(2)}, \ldots, r^{(100)}$ at random from $\{0, 1\}^n$.
    • Compute $A(Br^{(k)}) - Cr^{(k)}$ for each $k = 1, \ldots, 100$.
    • If $A(Br^{(k)}) - Cr^{(k)} = 0$ for all k, declare $AB = C$. Otherwise declare $AB \neq C$.
    Let E_k be the event that the algorithm gives the wrong answer at step k, i.e. $A(Br^{(k)}) = Cr^{(k)}$ but $AB \neq C$. We know that $P(E_k) \leq 1/2$ from Theorem 1.1. Let $E = \bigcap_{k=1}^{100} E_k$—this is the event where the algorithm gives $A(Br^{(k)}) = Cr^{(k)}$ for all k but $AB \neq C$. Since events E_k are independent, we have

$$P(E) = P(\bigcap_{k=1}^{100} E_k) = \prod_{k=1}^{100} P(E_k) \leq \left(\frac{1}{2}\right)^{100}.$$

2. The upper bound for the probability in the theorem will be $1/3$ instead of $1/2$. For the general case it will be $1/(d + 1)$.
3. ```
In [1]: using PyPlot
```

```
In [2]: function bound(D,d)
            count=0
            m=size(D)[1]
            for k in 1:10000
                u=ones(m)
                for j in 1:m
                    u[j]=rand(0:d)
                end
                if D*u==zeros(m)
                    count=count+1
                end
            end
```
```

```
                    end
                    count/10000
            end
```

Out[2]: bound (generic function with 1 method)

In [3]: D=rand(0:9,2,2)

Out[3]: 2×2 Array{Int64,2}:
         9 7
         5 9

In [4]: bound(D,1)

Out[4]: 0.2536

In [5]: bound(D,2)

Out[5]: 0.1149

In [6]: for i in 1:10
            D=rand(0:9,2,2)
            println(bound(D,1))
        end

```
0.2506
0.2546
0.2582
0.2458
0.2509
0.248
0.2576
0.2525
0.2458
0.2574
```

In [7]: mx=Array{Float64}(undef,0)
        for s in 2:10
            vals=Array{Float64}(undef,0)
            for i in 1:10
                D=rand(0:9,s,s)
                append!(vals,bound(D,1))
            end
            append!(mx,maximum(vals))
        end
        xvalues=2:1:10
        plot(xvalues,mx)

```
xlabel("Size of the matrix")
ylabel("Upper bound");
```

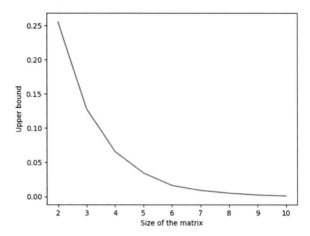

Project 3: A survey with three choices

$$P\{\text{"}yes\text{"}\} = P\{\text{"}yes\text{"} \mid Q1\}P\{Q1 \text{ is chosen}\}+$$
$$P\{\text{"}yes\text{"} \mid Q2\}P\{Q2 \text{ is chosen }\} + P\{\text{"}yes\text{"} \mid Q3\}P\{Q3 \text{ is chosen}\}$$

We have

$$P\{\text{"}yes\text{"}\} \rightarrow p$$
$$P\{\text{"}yes\text{"} \mid Q1 \} \rightarrow g$$
$$P\{Q1 \text{ is chosen }\} \rightarrow 4/6$$
$$P\{\text{"}yes\text{"} \mid Q2\} \rightarrow 1$$
$$P\{Q2 \text{ is chosen }\} \rightarrow 1/6$$
$$P\{\text{"}yes\text{"} \mid Q3\} \rightarrow 0$$
$$P\{Q3 \text{ is chosen }\} \rightarrow 1/6$$

Then

$$p = g\frac{4}{6} + 1 \cdot \frac{1}{6} + 0 \Rightarrow g = \frac{3}{2}p - \frac{1}{4}.$$

Project 4: The Haunting of Hill House

1. In [1]: functionsw_prob(n)

```
s=0
for n in 1:1000
    g=rand(1:3)
    if g==1
        w=0
    end
    if g==2
        w=1
    end
    if g==3
        w=1
    end
    s=s+w
end
s/1000
end
```

Out[1]: sw_prob (generic function with 1 method)

In [2]: sw_prob(1000)

Out[2]: 0.657

2. From Bayes' theorem,

$P(G_1|\text{Door 2 opened})$

$$= \frac{P(G_1)P(\text{Door 2 opened}|G_1)}{P(G_1)P(\text{Door 2 opened}|G_1) + P(G_2)P(\text{Door 2 opened}|G_2) + P(G_3)P(\text{Door 2 opened}|G_3)}.$$

Note that

$P(\text{Door 2 opened}|G_1) = 1/2, \quad P(\text{Door 2 opened}|G_2) = 0, \quad P(\text{Door 2 opened}|G_3) = 1.$

Also, $P(G_1) = P(G_2) = P(G_3) = 1/3$. Then

$$P(G_1|\text{Door 2 opened}) = 1/3.$$

Next we compute

$P(G_3|\text{Door 2 opened})$

$$= \frac{P(G_3)P(\text{Door 2 opened}|G_3)}{P(G_1)P(\text{Door 2 opened}|G_1) + P(G_2)P(\text{Door 2 opened}|G_2) + P(G_3)P(\text{Door 2 opened}|G_3)}$$

$$= 2/3.$$

Therefore, Mrs. Montague should switch from Door 1, her initial pick, to Door 3.
3. In this case the conditional probabilities are

$P(\text{Door 2 opened}|G_1) = 1/2, \quad P(\text{Door 2 opened}|G_2) = 1/2, \quad P(\text{Door 2 opened}|G_3) = 1/2$

and from Bayes' theorem

$P(G_1|\text{Door 2 opened}) = P(G_2|\text{Door 2 opened}) = P(G_3|\text{Door 2 opened}) = 1/3.$

The good ghost is not helpful in this scenario; the initial probabilities $P(G_1) = P(G_2) = P(G_3) = 1/3$ do not change with the extra information provided when Door 2 is opened.

Project 5: Benford's law

Here is the Julia code and the histograms. Load the packages PyPlot and JuliaDB before running this code.

```
In [1]: p(d)=log(10,1+1/d)
```

```
Out[1]: p (generic function with 1 method)
```

```
In [2]: data=loadtable("Iran_pres_2009.csv");
```

```
In [3]: res=select(data, :Mousavi);
        first_digs=[last(digits(n)) for n in res];
        first_digs_nz=filter(x->x>0,first_digs)
        xaxis=1:.1:9
        yvals=map(t->p(t),xaxis)
        plot(xaxis,yvals,label="Benford density")
        xlabel("Digits")
        ylabel("Relative frequency")
        plt.hist(first_digs_nz,9,density=true,
        label="Mousavi votes")
        legend(loc="upper center");
```

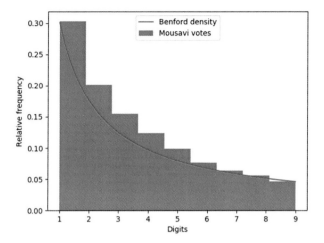

```
In [4]: res=select(data, :Ahmadinejad);
        first_digs=[last(digits(n)) for n in res];
        first_digs_nz=filter(x->x>0,first_digs)
        xaxis=1:.1:9
        yvals=map(t->p(t),xaxis)
        plot(xaxis,yvals,label="Benford density")
        xlabel("Digits")
        ylabel("Relative frequency")
        plt.hist(first_digs_nz,9,density=true,
        label="Ahmadinejad votes")
        legend(loc="upper center");
```

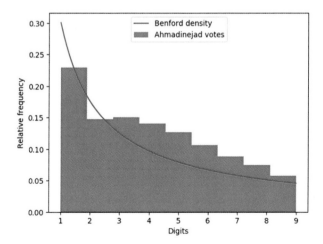

Project 6: Resurrect the Beetle!

2. Let X_1 be the number of boxes needed to get the first coupon. Clearly $X_1 = 1$ since each box has a coupon. Now let X_2 be the number of boxes needed to get the second distinct coupon, when we already have one coupon. Getting the second distinct coupon occurs with probability $\frac{n-1}{n}$—think of this as our "success" event. Then X_2 is a geometric random variable with probability of success $\frac{n-1}{n}$.

Similarly, let X_3 be the number of boxes needed to get the third distinct coupon, when we already have two types of coupons. Then $X_3 \sim Geo(\frac{n-2}{n})$. In general, $X_i \sim Geo(\frac{n-i+1}{n})$ and therefore $E[X_i] = \frac{n}{n-i+1}$.
Next observe that $X = \sum_{i=1}^{n} X_i$. Then

$$E[X] = \sum_{i=1}^{n} E[X_i] = 1 + \frac{n}{n-1} + \frac{n}{n-2} + \ldots + \frac{n}{2} + \frac{n}{1} = n\left(1 + \frac{1}{2} + \frac{1}{3} + \ldots + \frac{1}{n}\right).$$

3. $E[X] = 6(1 + 1/2 + 1/3 + 1/4 + 1/5 + 1/6) = 14.7$.
4.   a. The unknown number of distinct song elements, call it n, is the number of coupons. Let X be the number of observations needed to observe all the distinct song elements. We assume the bird sings each song element at random with equal probabilities.
  b. If $n = 625$, $E[X]$ can be computed using Julia as

```
In [1]: n=625
        s=0.
        for i in 1:n
            s=s+1/i
        end
        n*s
Out[1]: 4384.85443831541
```

Project 7: A professor's trick

1. Here we will prove parts 1. and 2. directly using mathematical induction. The induction basis $E[N_1] = \frac{n-1}{n-1} = 1$ is not too difficult to see. The induction hypothesis is $E[N_{k-1}] = \frac{n^{k-1}-1}{n-1}$. Now let's prove the statement for $E[N_k]$.

$$E[N_k] = \sum_{j=k-1}^{\infty} E[N_k \mid N_{k-1} = j]P\{N_{k-1} = j\}$$

$$= \sum_{j=k-1}^{\infty} \left(\frac{1}{n}(j+1) + \frac{n-1}{n}(j + E[N_k]) \right) P\{N_{k-1} = j\}$$

$$= \frac{1}{n} \sum_{j=k-1}^{\infty} (j+1)P\{N_{k-1} = j\} + \frac{n-1}{n} \sum_{j=k-1}^{\infty} (j + E[N_k])P\{N_{k-1} = j\}$$

$$= \frac{1}{n} \left(\sum_{j=k-1}^{\infty} jP\{N_{k-1} = j\} + \sum_{j=k-1}^{\infty} P\{N_{k-1} = j\} \right) + \frac{n-1}{n} \sum_{j=k-1}^{\infty} jP\{N_{k-1} = j\}$$

$$+ \frac{n-1}{n}E[N_k] \sum_{j=k-1}^{\infty} P\{N_{k-1} = j\}$$

$$= \frac{1}{n}(E[N_{k-1}] + 1) + \frac{n-1}{n}E[N_{k-1}] + \frac{n-1}{n}E[N_k].$$

Solving this equation for $E[N_k]$ we get

$$\frac{E[N_k]}{n} = \frac{1}{n}(E[N_{k-1}] + 1 + (n-1)E[N_{k-1}])$$

$$\Rightarrow E[N_k] = E[N_{k-1}] + 1 + (n-1)E[N_{k-1}] = nE[N_{k-1}] + 1.$$

Now we use induction hypothesis that $E[N_{k-1}] = \frac{n^{k-1}-1}{n-1}$ to obtain

$$E[N_k] = n \left(\frac{n^{k-1} - 1}{n-1} \right) + 1 = \frac{n^k - 1}{n-1}.$$

2. $P\{N_k \geq a\} \leq \frac{(n^k-1)}{a(n-1)}$.
3. In the case of coin flips $n = 2$, the professor uses $k = 4$ so $E[N_4] = 2^4 - 1 = 15$. So the average number of flips to observe 4 consecutive heads or tails is 15. The students flip the coin 100 times. From part 2, $P\{N_4 \geq 100\} \leq 15/100 = 0.15$. Therefore, the probability that a student does not observe 4 consecutive heads or tails in 100 flips is less than 0.15. Since this probability is rather small, the professor guesses if a student does not have 4 consecutive heads or tails in 100 flips, then it is likely due to trying to guess the outcomes instead of actually flipping them.

Why does the professor say she is 95% confident that at least one out of 4 students did not actually flip the coin? When a student observes 4 consecutive heads or tails in 100 flips, let's label the event as "success", and otherwise "failure". Let's use the upper bound 0.15 as the failure probability. If all students were actually flipping the coin, what would be the probability that there are 6 successes and 4 failures? The answer is $\binom{10}{6}(0.85)^6(0.15)^4 = 0.04$, which is about 5%. That is why the professor is 95% certain that observing 6 successes (and 4 failures) is the result of foul play, and she guesses the tricksters are among the "failures".

Project 8: Monte Carlo integration

Hit-or-miss Monte Carlo

Solutions to problem 2:

1. $P\{X_i = 1\} = I$.
2. $S = \sum_{i=1}^{N} X_i$. S is a binomial distribution with parameters N and probability of success I.
3. $E[S] = NI$, and thus $E[S/N] = I$.
4. $Var(\frac{S}{N}) = \frac{1}{N^2} Var(S) = \frac{1}{N^2} I(1 - I)N = \frac{I(1-I)}{N}$.

Sample mean Monte Carlo

1. In [1]:
```
function mc(N)
    sum=0.
    for i in 1:N
        sum=sum+exp((rand()^2))
    end
    return sum/N
end
```

Out[1]: mc (generic function with 1 method)

In [2]: mc(1000)

Out[2]: 1.4716427305948323

In [3]: mc(10000)

Out[3]: 1.461656048504788

Let's also code the basic Simpson's quadrature to compare with Monte Carlo.

In [4]:
```
function simpson(f::Function,a,b)
    h=(b-a)/2
    h*(f(a)+4f(a+h)+f(b))/3
end
```

```
Out[4]: simpson (generic function with 1 method)
```

```
In [5]: simpson(x->exp((x^2)),0,1)
```

```
Out[5]: 1.4757305825350018
```

2. Next problem is $\int_0^1 \int_0^1 e^{(x+y)^2} dxdy$.

```
In [6]: function mct(N)
            sum=0.
            for i in 1:N
                sum=sum+exp((rand()+rand())^2)
            end
            return sum/N
        end
```

```
Out[6]: mct (generic function with 1 method)
```

```
In [7]: mct(1000)
```

```
Out[7]: 4.6792007051291495
```

```
In [8]: mct(10000)
```

```
Out[8]: 4.8817082581602005
```

WolframAlpha's approximation for the above integral is 4.89916.

3. This is for the integral of any function defined on (a, b).

```
In [9]: function mc(g::Function,a,b,N)
            sum=0.
            for i in 1:N
                sum=sum+g(a+(b-a)*rand())*(b-a)
            end
            return sum/N
        end
```

```
Out[9]: mc (generic function with 2 methods)
```

4. Note that

$$Var\left(\frac{1}{N}\sum_{i=1}^{N} g(U_i)\right) = \frac{1}{N}\left(\int g^2(x)dx - I^2\right)$$

and from part 1

$$Var(S/N) = \frac{1}{N}I(1-I).$$

Then it suffices to prove $\int g^2(x)dx \leq I = \int g(x)dx$. This statement is true since $0 < g^2(x) \leq g(x)$, which follows from $0 < g(x) < 1$.

Project 9: Florida Panther

The following data is from https://myfwc.com/wildlifehabitats/wildlife/panther/pulse/:

Year	2018	2017	2016	2015	2014
Births	9	19	14	15	32
Deaths	30	30	42	42	34

In [1]: `using PyPlot`

In [2]: `using Statistics`

In [3]: `b=[9,19,14,15,32]`
        `d=[30,30,42,42,34];`

In [4]: `brate=b/120`

Out[4]: `5-element Array{Float64,1}:`
        ` 0.075`
        ` 0.15833333333333333`
        ` 0.11666666666666667`
        ` 0.125`
        ` 0.26666666666666666`

In [5]: `mean(brate)`

Out[5]: `0.14833333333333334`

In [6]: `std(brate)`

Out[6]: `0.07250478911385402`

In [7]: `drate=d/120`

Out[7]: `5-element Array{Float64,1}:`
        ` 0.25`
        ` 0.25`
        ` 0.35`
        ` 0.35`
        ` 0.2833333333333333`

In [8]: `mean(drate)`

```
Out[8]:  0.2966666666666667

In [9]:  std(drate)

Out[9]:  0.05055250296034366

In [10]:  function x(n)
                b=0.15+0.07*randn()
                d=0.3+0.05*randn()
                if n==1
                    return 120
                else
                (b-d+1)*x(n-1)
                end
         end

Out[10]:  x (generic function with 1 method)

In [11]:  pop=[x(n) for n=1:15]

Out[11]:  15-element Array{Real,1}:
            120
            114.87059905494884
             91.05063601079928
             79.27207001148582
             87.74996151202089
             48.66807597422959
             21.624759843163503
             34.40825544710225
             25.99836370154169
             25.67660431883496
             25.41084548077546
             15.670946562373858
             16.989603998898662
              7.366857480813708
             14.435932040482387

In [17]:  xvalues=1:1:15
          plot(xvalues,pop)
          xlabel("Year")
          ylabel("Population");
```

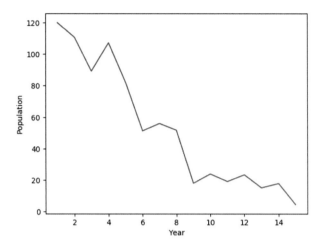

```
In [13]: function extinct(m)
             count=0
             for i in 1:m
                 pop=[x(n) for n=1:15]
                 if minimum(pop)<10
                     count=count+1
                 end
             end
             println("The probability of extinction is ", count/m)
         end

Out[13]: extinct (generic function with 1 method)

In [14]: extinct(10000)

The probability of extinction is 0.5567
```

Project 10: Can humans generate random numbers?

This is the application of the test to Julia's random numbers.

```
In [1]: using Distributions

In [2]: n=500
        lsum=0.
```

```
        digs=rand(0:9,n)
        freq=Array{Float64}(undef,0)
        for i in 1:10
            append!(freq,count(n->n==i-1,digs))
        end
        for i in 1:10
            lsum=lsum+10*(freq[i]-n/10)^2/n
        end
        return lsum
```

Out[2]: **10.16**

In [3]: **X=Chisq(9)**
        1-cdf.(X,10.16)

Out[3]: **0.33768835649023954**

The digits pass the χ^2-test.

In [4]: **n=500**
        m=250
        lsum=0.
        digs=rand(0:9,n)
        pairs=[(digs[2*i-1],digs[2*i]) for i in 1:m]
        freq=Array{Float64}(undef,0)
        for i in 1:10
            append!(freq,count(n->n==(i-1,i-1),pairs))
        end
        for i in 1:10
            lsum=lsum+100*(freq[i]-m/100)^2/m
        end
        rfreq=m-sum(freq)
        lsum=lsum+10*(rfreq-9*m/10)^2/(9*m)
        return lsum

Out[4]: **11.404444444444444**

In [5]: **X=Chisq(10)**
        1-cdf.(X,11.4)

Out[5]: **0.3272148022285587**

The pairs of digits pass the χ^2-test as well.

Project 11: Analyzing a die game with Markov chains

Labeling the rows as $1', 1, 2, 2'$ (top to bottom), and labeling the columns similarly (left to right), we obtain the following transition matrix:

$$P = \begin{bmatrix} 1 & 0 & 0 & 0 \\ 1/6 & 1/3 & 1/2 & 0 \\ 0 & 1/2 & 1/3 & 1/6 \\ 0 & 0 & 0 & 1 \end{bmatrix}.$$

The n-step transition probability $p_{11'}(n)$ is the probability that player 1 starts the game and wins it in n steps. Computing higher powers with Julia, we estimate the limit as

$$P^\infty = \begin{bmatrix} 1 & 0 & 0 & 0 \\ 0.57 & 0 & 0 & 0.43 \\ 0.43 & 0 & 0 & 0.57 \\ 0 & 0 & 0 & 1 \end{bmatrix}.$$

Therefore the probability that the player who starts the game wins is 0.57.

Project 12: Market share of shoe brands

We first enter the number of transitions data given in Appendix B in Julia:

```
In [1]: P=[5 7 2 1; 5 10 4 2; 8 2 3 3; 15 15 8 10]
```

```
Out[1]: 4×4 Array{Int64,2}:
          5    7   2    1
          5   10   4    2
          8    2   3    3
         15   15   8   10
```

To compute the transition probabilities, we need to divide each row of P by the row sum:

```
In [2]: Pt=[5/157/152/15 1/15; 5/21 10/21 4/21 2/21; 8/16 2/163/163/16;
            15/4815/488/4810/48]
```

```
Out[2]: 4×4 Array{Float64,2}:
         0.333333   0.466667   0.133333   0.0666667
         0.238095   0.47619    0.190476   0.0952381
         0.5        0.125      0.1875     0.1875
         0.3125     0.3125     0.166667   0.208333
```

The steady-state vector π is the solution of $A\pi = c$ where A, c are given as (using two decimal digits)

```
In [3]: A=[1 1 1 1; -0.67 0.24 0.5 0.31; 0.13 0.19 -0.81 0.17;
            0.07 0.095 0.19 -0.79]
        c=[1;0;0;0];
```

We solve this equation for π next:

```
In [4]: A\c
```

```
Out[4]: 4-element Array{Float64,1}:
        0.3208083810505234
        0.39440566578847597
        0.16842426334055977
        0.11636168982044079
```

Therefore Adidas will pay 32% of the yearly maintenance of $10,000, Nike will pay 39% of it, etc.

Project 13: Modeling insect movement

1.
```
In [1]: using PyPlot

In [2]: function rwalk(n)
            k=10
            x=zeros(n)
            y=zeros(n)
            for i in 2 : n
                z1=randn()
                z2=randn()
                x[i]=x[i-1]+z1
                if x[i]>k
                    x[i]=k
                end
                if x[i]<-k
                    x[i]=-k
                end
                y[i]=y[i-1]+z2
                if y[i]>k
                    y[i]=k
                end
                if y[i]<-k
                    y[i]=-k
                end
```

```
            end
            plot(x,y)
            xlim(-k,k)
            ylim(-k,k)
        end
```

Out[2]: rwalk (generic function with 1 method)

In [3]: rwalk(400);

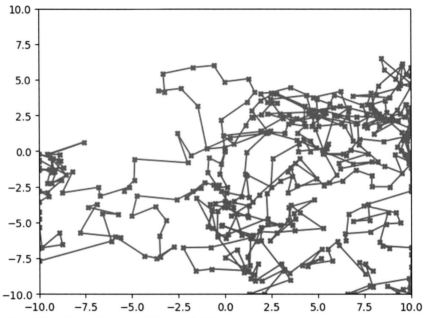

2. In [4]: function rwalklength(n)

```
            k=10
            x=zeros(n)
            y=zeros(n)
            length=0
            lengthsq=0
            for i in 2 : n
                z1=randn()
                z2=randn()
                x[i]=x[i-1]+z1
                if x[i]>k
                    x[i]=k
                end
                if x[i]<-k
                    x[i]=-k
                end
```

```
            y[i]=y[i-1]+z2
            if y[i]>k
                y[i]=k
            end
            if y[i]<-k
                y[i]=-k
            end
            length=length+(z1^2+z2^2)^0.5
            lengthsq=lengthsq+(z1^2+z2^2)
        end
        println("average length $(length/n)")
        println("average length squared $(lengthsq/n)")
    end
```

Out[4]: rwalklength (generic function with 1 method)

In [5]: rwalklength(1000)

average length 1.2268474579427824
average length squared 1.916879768077854

In [6]: rwalklength(100000)

average length 1.2524559977662268
average length squared 1.9992183596107311

3. First observe that the length of the ith move is $l(i) = \sqrt{z_1^2 + z_2^2}$ where z_1, z_2 are the random numbers from the standard normal distribution that were used in computing

$$x(i) = x(i - 1) + z_1$$
$$y(i) = y(i - 1) + z_2.$$

The notation $l(i)$ is for the specific length in one trajectory, and $L(i)$ for the random variable whose values are $l(i)$. Then

$$E[L(i)^2] = E[Z_1^2] + E[Z_2^2] = Var(Z_1) + Var(Z_2) = 2,$$

where Z_1, Z_2 are independent standard normal random variables.

4. We can ensure the insect only moves toward the top-right corner once it enters the rectangle $[5, 10] \times [5, 10]$ by replacing z_1, z_2 in Eq. (5.2) with their absolute values, $|z_1|, |z_2|$, when the insect is in the rectangle.

In [7]: function dirwalk()
            k=10

```
x=[0.]
y=[0.]
while last(x)<k/2 || last(y)<k/2
    z1=randn()
    z2=randn()
    xn=last(x)+z1
    if xn>k
        xn=k
    end
    if xn<=-k
        xn=-k
    end
    append!(x,xn)
    yn=last(y)+z2
    if yn>k
        yn=k
    end
    if yn<-k
        yn=-k
    end
    append!(y,yn)
end
while last(x)<k || last(y)<k
    z1=randn()
    z2=randn()
    if last(x)+abs(z1)>k
        xn=k
    else
        xn=last(x)+abs(z1)
    end
    append!(x,xn)
    if last(y)+abs(z2)>k
        yn=k
    else
        yn=last(y)+abs(z2)
    end
    append!(y,yn)
end
plot(x,y,marker="X",markersize=5)
xlim(-k,k)
ylim(-k,k)
hlines(5,5,10)
vlines(5,5,10)
end
```

Out[7]: **dirwalk (generic function with 1 method)**

In [8]: **dirwalk()**

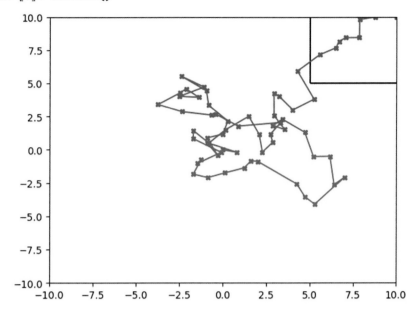

Out[8]: **PyObject <matplotlib.collections.LineCollection object at 0x11ab348d0>**

5. In [9]: **function movestofood(m)**
   ```
                 k=10
                 count=0
                 for i in 1:m
                     x=[0.]
                     y=[0.]
                     while last(x)<k/2 || last(y)<k/2
                         z1=randn()
                         z2=randn()
                         xn=last(x)+z1
                         if xn>k
                             xn=k
                         end
                         if xn<=-k
                             xn=-k
                         end
                         append!(x,xn)
                         yn=last(y)+z2
                         if yn>k
                             yn=k
                         end
   ```

```
                    if yn<-k
                        yn=-k
                    end
                    append!(y,yn)
                end
                while last(x)<k || last(y)<k
                    z1=randn()
                    z2=randn()
                    if last(x)+abs(z1)>k
                        xn=k
                    else
                        xn=last(x)+abs(z1)
                    end
                    append!(x,xn)
                    if last(y)+abs(z2)>k
                        yn=k
                    else
                        yn=last(y)+abs(z2)
                    end
                    append!(y,yn)
                end
                count=count+length(x)
            end
            println("The avg number of moves till food is $(count/m)")
        end
```

Out[9]: movestofood (generic function with 1 method)

In [10]: movestofood(1000)

The avg number of moves till food is 468.073

In [11]: movestofood(10000)

The avg number of moves till food is 437.9129

Project 14: Option pricing

1. The data used in this solution, obtained from https://finance.yahoo.com/quote/
   GE/options?p=GE, had the following parameters: the current date ($t_0 = 0$) is
   October 8, 2019, the expiry is $T = 10/252$ (the option expires in 10 days), the
   strike price is $K = 4.5$, the last price the option was traded for is $\tilde{C} = 4.15$,
   and the last price the stock was traded for is $S_0 = 8.27$. The interest rate for a
   one-year certificate of deposit is $r = 2.25\%$.

2. The sample standard deviation of $\log \frac{S(t_i)}{S(t_{i-1})}$ for the GE stock during 10/08/2018–10/08/2019 is 0.013, where $S(t_i)$ are the daily opening prices[1]. Then $\sigma = \sqrt{252}(0.013) = 0.21$.

3. In [6]: `function calloptionprice (r,sigma,szero,K,T,N)`

```
        sum=0
        for j in 1:N
        sfinal=szero*exp((r-sigma^2/2)*T+sigma*sqrt(T)*randn())
        sum=sum+max(sfinal-K,0)
        end
        exp(-r*T)*sum/N
   end
```

Out[6]: `calloptionprice (generic function with 1 method)`

In [7]: `calloptionprice(0.0225,0.21,8.27,4.5,10/252,10000)`

Out[7]: `3.774381865391275`

---

[1] Data downloaded from https://finance.yahoo.com/quote/GE/

References

[1] Aoyama, H., Fujiwara, Y., Ikeda, Y., Iyetomi, H., & Souma, W. (2010). *Econophysics and companies: Statistical life and death in complex business networks.* Cambridge University Press.

[2] Balsby, T. J. S., & Hansen, P. (2010). Element repertoire: Change and development with age in Whitethroat Sylvia communis song. *Journal of Ornithology, 151*(2), 469–476.

[3] Bayes, T. (1763). An essay towards solving a problem in the doctrine of chances. By the late Rev. Mr. Bayes, FRS communicated by Mr. Price, in a letter to John Canton, AMFR S. *Philosophical Transactions of the Royal Society of London,* (53), 370–418.

[4] Benford, F. (1938). The law of anomalous numbers. *Proceedings of the American Philosophical Society, 78*(4), 551–572.

[5] Billingsley, P. (2008). *Probability and measure.* Wiley.

[6] Brown, R. (1828). A brief account of microscopical observations made in the months of June, July and August 1827, on the particles contained in the pollen of plants; and on the general existence of active molecules in organic and inorganic bodies. *The Philosophical Magazine, 4*(21), 161–173.

[7] Figurska, M., Stańczyk, M., & Kulesza, K. (2008). Humans cannot consciously generate random numbers sequences: Polemic study. *Medical Hypotheses, 70*(1), 182–185.

[8] Freivalds, R. (1977). Probabilistic machines can use less running time. In *IFIP Congress 1977* (pp. 839–842).

[9] Fujiwara, Y. (2004). Zipf law in firms bankruptcy. *Physica A: Statistical Mechanics and Its Applications, 337*(1–2), 219–230.

[10] Greenberg, B., Abul-Ela, A., Simmons, W., & Horvitz, D. (1969). The unrelated question randomized response model: Theoretical framework. *Journal of the American Statistical Association, 64*(326), 520–539.

[11] Jackson, S. (2013). *The haunting of hill house.* Penguin Classics.

[12] Lawrance, A. E. (1969). Playing with probability. *The Mathematical Gazette, 53*(386), 347–354.

© The Editor(s) (if applicable) and The Author(s), under exclusive license to Springer Nature Switzerland AG 2020
G. Ökten, *Probability and Simulation*, Springer Undergraduate Texts in Mathematics and Technology, https://doi.org/10.1007/978-3-030-56070-6

[13] Maull, W., & Berry, J. (1997). Modelling the coupon collector's problem. *Teaching Statistics*, *19*(2), 43–46.

[14] Kahneman, D. (2011). *Thinking, fast and slow*. Macmillan.

[15] Kareiva, P. M., & Shigesada, N. (1983). Analyzing insect movement as a correlated random walk. *Oecologia*, *56*(2–3), 234–238.

[16] Markov, A. A. (2006). An example of statistical investigation of the text Eugene Onegin concerning the connection of samples in chains. *Science in Context*, *19*(4), 591–600. (Translation of "An Example of Statistical Investigation of the Text Eugene Onegin Concerning the Connection of Samples in Chains", A. A. Markov, 1913.)

[17] Newcomb, S. (1881). Note on the frequency of use of the different digits in natural numbers. *American Journal of Mathematics*, *4*(1/4), 39–40.

[18] Nigrini, M. J. (2012). *Benford's Law: Applications for forensic accounting, auditing, and fraud detection* (Vol. 586). Wiley.

[19] Ökten, G. (2019). *First semester in numerical analysis with Julia*. Florida State University Libraries, https://doi.org/10.33009/jul.

[20] Pearson, K. (1900). On the criterion that a given system of deviations from the probable in the case of a correlated system of variables is such that it can be reasonably supposed to have arisen from random sampling. *The London, Edinburgh, and Dublin Philosophical Magazine and Journal of Science*, *50*(302), 157–175.

[21] Persaud, N. (2005). Humans can consciously generate random number sequences: A possible test for artificial intelligence. *Medical Hypotheses*, *65*(2), 211–214.

[22] Samuelson, P. A. (1965). Proof that properly anticipated prices fluctuate randomly. *Management Review*, *6*(2).

[23] Stewart, I. (1996). Monopoly revisited. *Scientific American*, *275*(4), 116–119.

[24] Quine, M. P., & Seneta, E. (1987). Bortkiewicz's data and the law of small numbers. *International Statistical Review/Revue Internationale De Statistique*, *55*(2), 173–181.

[25] Warner, S. L. (1965). Randomized response: A survey technique for eliminating evasive answer bias. *Journal of the American Statistical Association*, *60*(309), 63–69.

[26] Welch, R. E., & Frick, T. W. (1993). Computerized adaptive testing in instructional settings. *Educational Technology Research and Development*, *41*(3), 47–62.

Index

© The Editor(s) (if applicable) and The Author(s), under exclusive
license to Springer Nature Switzerland AG 2020
G. Ökten, *Probability and Simulation*, Springer Undergraduate Texts
in Mathematics and Technology, https://doi.org/10.1007/978-3-030-56070-6

Printed in the United States
By Bookmasters